The
Relativity of
Deviance

The Relativity of Deviance

John Curra

Sage Publications
International Educational and Professional Publisher
Thousand Oaks   London   New Delhi

For information:

SAGE Publications, Inc.
2455 Teller Road
Thousand Oaks, California 91320
E-mail: order@sagepub.com

SAGE Publications Ltd.
6 Bonhill Street
London EC2A 4PU
United Kingdom

SAGE Publications India Pvt. Ltd.
M-32 Market
Greater Kailash I
New Delhi 110 048 India

Printed in the United States of America

Library of Congress Cataloging-in-Publication Data

Curra, John.
    The relativity of deviance / by John Curra.
      p.   cm.
    Includes bibliographical references and index.
    ISBN 0-7619-0777-7 (cloth: alk. paper)
    ISBN 0-7619-0778-5 (pbk.: alk. paper)
    1. Deviant behavior.
    HM811.C87   1999
    302.5′ 42—dc21   99-006305

This book is printed on acid-free paper.

00   01   02   03   04   05   10   9   8   7   6   5   4   3   2   1

Acquiring Editor: Peter Labella
Editorial Assistant: Reneé Piernot
Production Editor: Wendy Westgate
Production Assistant: Patricia Zeman
Designer/Typesetter: Janelle LeMaster
Cover Designer: Michelle Lee
Indexer: Molly Hall

Contents

Preface

Human actions and attributes do not, of course, exist in a social vacuum; they are always of great interest and concern to others. Depending on time, place, situation, the particular temperament of the individuals involved, and a host of other factors, some activities and attributes are praised, some are condemned, and others are ignored, at least for a time. Readers of this book will see that what qualifies as deviance varies from place to place, time to time, and situation to situation. They will see that deviance cannot be understood apart from its social setting, and they will be introduced to many examples to show that this is true. What exists could almost always be otherwise, and a change in social settings and interpersonal relationships almost always produces a change in the nature of deviance.

I have written this book to be appealing to anyone who is fascinated by the great diversity in human experience and the even greater diversity in social representations of it. The book will help readers to appreciate that the meanings of deviance are problematic and always changing. They will come to realize that no intrinsic or objective qualities exist to separate deviance from nondeviance and that deviance is far from absolute and uniform. This book will encourage readers to evaluate the understandings of deviance that they have acquired throughout their lives, and it will prompt them to think about their own definitions of deviance.

The book should be of value as a teaching aid in two situations. It can be used as a supplemental reading in courses in sociology, anthropology, criminology, juvenile delinquency, psychology, law enforcement, or social work in which instructors want their students to have a clearer sense of the relative nature of social deviance than is offered by most books. I also think that for those instructors who want their students to read about deviance but who do not wish to use a standard textbook, this book offers an alternative. It drives home the point that nothing is inherently deviant and that what exists could always be different.

I have carefully selected the examples of differentness and/or deviance that are found throughout the book. In the following pages, readers will meet the blue people of Kentucky, a woman who believes that she is a vampire, the thugs of India, autoerotic asphyxiates, berdaches, mutilators, and a man who ate an entire airplane. Some of the examples are funny; some are sad; some are shocking. The examples are more than journalistic reports of quirky events, however. They are included to show again and again that deviance is a social construction of individuals in particular situations. I have used all the book's examples in my classes with great success; I hope that they work as well for others.

There are many things that the book is not, too numerous to mention. However, it must be stressed that it is not a textbook. A reader will not find formal discussions of labeling theory, radical deviance theory, ethnomethodology, control theory, or anomie theory; nor will he or she find discussions of the measurement and the control of deviance. It is true that the only acceptable way to understand deviance is to use variables from sociological theories, such as power, conflict, norms, labeling, subculture, motive, intent, retrospective interpretation, stigma, accounts, ideologies, and social reaction, so these concepts do have a prominent place in the following pages. However, minor attention will be devoted to theoretical developments. I am more concerned with which concepts help us understand human diversity, social deviance, and relativity the best than I am with which theory is the best.

Chapter 1 shows how deviance can be understood as a social event by discussing concepts such as egocentricity, ethnocentricity, symbols, ideal culture, norms, sanctions, and ideologies. This chapter will show the student of deviance that we live in a world of multiple realities (even though one particular reality seems real and concrete to those people who cherish it) and that deviance must be understood as situational and problematic. Readers will come to understand what relativity means and how the term can help to clarify the relationship between human diversity and social deviance.

Each of the book's remaining chapters, in addition to discussing the chapter's central topic, discusses a different facet of the relativity of deviance. Chapter 2 (dealing with appearance, identities, and stigma) shows that people can be deviant for things over which they have little control, and it insists that understanding and explaining deviance require a language of relationships, not of behavior or of attributes themselves. To show that this is true, the chapter discusses physical appearance and its social construction, bodily adornment such as tattooing, and eccentrics (and their eccentricities). Chapter 3 deals with sex, gender, and sexuality. It shows that sexual diversity—rooted in biology, psychology, culture, and society—is transformed into sexual deviance whenever humans and what they do, think, or feel are pigeonholed and then devalued. Few universals exist, and sexuality is a complex and ever-changing human relationship. Chapter 4 examines predatory violence. It shows that even serious acts that produce great harm are not necessarily deviant at all times and in all places and that even a form of human behavior that is usually viewed as intrinsically deviant really is not. Chapter 5 deals with sexual violence. It shows that violence is often normative and produced by social conflicts. However, depending on time and place, not all social conflicts get transformed into social deviance. The discussion of suicide, the topic of Chapter 6, will present readers with information about the framing of social events. Readers will see that even a generally disapproved-of event such as suicide is a socially constructed category with diverse meanings and that it, too, is not condemned in all times and places. Chapter 7 discusses the taking of other people's belongings. It shows the influential role played by power in the social construction of deviance. Some people's taking is considered perfectly all right, whereas other people's taking is criminalized and punished to the fullest extent of the law. Chapter 8 discusses drugs and deviance. Readers will see that some groups use their power to embark on moral crusades to brand other people's drug use as sick and wholly unacceptable. The final chapter shows how human diversity, social conflicts, power, organizational interests, social control, and naming and classifying all work together to construct that social deviance known as mental disorder.

My overriding goal has been to write a lively and provocative book for anyone who is interested in social deviance. I have tried to give readers a sense of the social and relative nature of deviance by examining sex, violence, theft, drugs, and mental disorders in such a way as to show that what is permissible in one setting is not permissible in another. Readers will be shown why it is so valuable and so necessary for them to maintain a relativistic view in their own understandings and explanations of social deviance. The book contains what I hope are interesting and thought-provoking examples of deviance, designed to show readers the great array in human experience and the even greater array of typifications of it. Fundamentally, I believe that this book will encourage a reader to reconsider his or her particular view of what is deviant and what is not.

The Dynamic Nature of Deviance

Introduction: Understanding Social Deviance

Social deviance is a persistent and common feature of societies, communities, and groups. Whenever people get together, some of them seem to hurt, annoy, or unsettle others. This does not mean, however, that deviance is some personal imperfection. *People acting together* create social deviance by what they believe, feel, say, and do, and we will find deviance and deviants almost everywhere that we go. Nothing can be deviant in a social vacuum, and definitions and social reactions strongly affect the types of deviance that exist in societies (Schur, 1971, pp. 16-17). The study of social deviance is more than a study of a type of behavior or of an individual attribute; it is a study of social relationships and socially constructed *perspectives* on human behavior and human individuals (Goffman, 1961). Good and bad are mutually defined, and each of them has meaning only in terms of the other (Margolin, 1993, p. 511). We must always know what is made of an act or attribute socially.

Deviance changes from society to society, and it changes in any given society over time. At one time or place, for some people, drinking alcohol is perfectly proper, and at another time and place it is forbidden; at one time or place, smoking cigarettes may be a sign of maturity and sophistication, and at some other time or place, it is a sign of immaturity and irresponsibility. The deviancy of some action or attribute cannot be determined simply by examining it closely and objectively. No human act or attribute will be universally judged as improper by people in all societies—large and small, industrial and nonindustrial—at all times. Human beings are simply too inventive in assigning positive and negative labels to the many things that they do for us to believe that deviance is separable from nondeviance because of some inherent, intrinsic, or objective quality of acts or attributes. Social deviance exists because some groups judge and evaluate what other groups are doing (Matza, 1969, pp. 41-53).

Though we must be concerned with human diversity and come to some understanding of why people in one setting are different from people in another setting, this is not enough. We must also understand that some people are responsible for constructing claims or understandings about what they find troubling or upsetting in other people (Spector & Kitsuse, 1977). Drawing these social lines makes some differences between groups look bigger than they are and some similarities look smaller than they are. These social divisions, artificial as they are, categorize and segregate, so they tell beholders what to recognize and what to ignore (Zerubavel, 1991). We must examine the content of these conceptions of deviance and understand why some people created them in the first place. Because we seek to understand deviance as a social relationship, we cannot only examine the diversity of human behavior; we must also examine the diversity of claims, labels, or definitions of it.

Being Centered: Egocentricity
and Ethnocentricity

As children, we were all *egocentric* (Piaget, 1948). We each lived in our own little world, and we were not yet aware of, or concerned with, the viewpoint of others. We could be aware of other people, but we had much more difficulty imagining how we appeared to them. This egocentricity colored everything that we did and everything that we believed and felt. We may have realized that we had parents, but we were less sure that our parents were in charge of us; we may have known that we had brothers and sisters, but we were less sure that our brothers and sisters had *us* as brothers or sisters. It was only through a great deal of interaction with others that we reached the point where we came to see ourselves through their eyes, and we wondered more and more about their thoughts and feelings.

The egocentricity of childhood colored our attitudes about right and wrong, good and bad, proper and improper, and correct and incorrect. When we were children, we decided what should or should not be done in a very selfish and direct way: If it brought us pleasure, it was "right," and if it brought us displeasure, it was "wrong." Others may not have liked what we did, and they took time to correct us, but in the initial period of social development, egocentricity was the rule. Most of us tried to maximize our own personal pleasure and minimize our own personal displeasure even if our selfishness was irritating or upsetting to others.

Socialization is a continuing process of acquiring the fundamentals necessary for group living and learning the heritage of a society. As individuals are socialized, they internalize the perspectives and expectations of others, and they become more conscious of, and potentially more sympathetic toward, others' outlooks and interests. Fully socialized individuals are able to take the roles of others, and they may refuse to do anything that would hurt others even if doing so would bring them some immediate advantage.

The internalization of a culture (or at least parts of it) is an important goal of socialization. Culture is a system of designs for living or shared understandings that members of a society use as they act together (Kluckhohn, 1949). These designs or understandings are created by people at some place and at some time, and they are then transmitted from group to group or generation to generation. People in a society (or a part of it) find that certain ways of acting, thinking, or feeling seem better than other ways, and these designs for living are then encouraged or even demanded. "Do unto others as you would have them do unto you" is part of U.S. culture, as is the injunction to "obey all traffic laws." If individuals do internalize a culture, obey its rules, and honor its values, they are usually defined by other members of that society as civilized and responsible members of the group.

Socialization alters the egocentricity of childhood. As we become more aware of others, we are forced to start looking at our society and the people in it in new ways. We eventually come to think, feel, and act in *anticipation* of the impact we will have on others. This continual movement away from egocentricity makes it less likely that we will intentionally injure someone else, but it does not automatically guarantee it. Though the centeredness of childhood is modified by the socialization process, centeredness does not end. People will maintain some of their egocentricity throughout their lives (Kumbasar, Romney, & Batchelder, 1994, p. 499). They will continue to do what is good for them *as individuals* even if it is contrary to the wishes and interests of others. This centeredness may create a type of tunnel vision in which individuals do not consider the effects of their activities on others, increasing substantially the likelihood that what they do or say will injure or upset someone else (Sheley & Corsino, 1994, p. 269).

The process of socialization itself creates another kind of centeredness called *ethnocentrism.* Socialization encourages people to believe that their society's values, standards, and customs are better than the values, standards, and customs of people in other societies. It is easy to see how ethnocentrism could lead to a vilification of members of some out-group. If Americans were to decide that they were better than, say, Martians because Americans played football better, this could easily be a reflection of ethnocentrism. A custom in one culture, football, would be used to evaluate and then denigrate members of some other culture who had no knowledge of football. Using the values, standards, or customs of one culture to evaluate people of some other culture will usually produce a great deal of distortion and misunderstanding. Customs and belief systems must be understood in the context of the cultural system of which they are a part.

Egocentricity and ethnocentrism—individual- and group-based centeredness—have important consequences for what we do and how we view each other. Egocentricity and ethnocentrism make it easy for some people in some cultures to devalue ways of acting, feeling, and thinking different from their own. Egocentricity and ethnocentrism make it more likely that some people will define as inferior other people who are different from them or who do things differently from what they do. Difference is easily transformed into deviance, and deviance is easily transformed into abnormality.

We must remember that egocentricity and ethnocentrism are biases. They represent ways of thinking, feeling, and acting that are conditioned by an *overidentification* with one's own individual interests or one's own culture. Claims about normality or abnormality, health or illness, morality or immorality, conformity or deviance *always* reflect some form of centeredness because these claims are culturally bound and historically specific. The assessment that something is abnormal, sick, immoral, or deviant always reflects the interests or preferences of some specific group in some particular society at some particular point in time. Changes in people, places, or times produce changes in definitions of normal and abnormal.

The Symbolic Organization of Human Experience

A *symbol* stands for or represents a person, object, situation, or experience because of group agreement and learning. Symbols are very flexible and can be used to evoke images of the thing symbolized even if that thing is not physically present or is abstract and never had physical substance. Practically any sound or physical gesture could be a symbol if enough people understood what the symbol represented and agreed to use it. The word *book* is a symbol of this object you are holding in your hands, and the period at the end of this sentence symbolizes that you are supposed to stop briefly when you reach it. Symbols are always learned, and they form the basis of language systems.

Symbols and the language systems of which they are a part provide a universe of discourse and meaning for humans. A major part of human development is the learning of the symbols found in one's culture; learning the symbols helps the infant become a full-fledged member of a society. Humans do not live only in a physical world of smells, sights, sounds, tastes, and touches. They live in a world of "hot and humid," "party dresses," "frozen pepperoni pizzas," "governments by the people and for the people," and many other things, material and nonmaterial, that are identified and symbolized in a culture.

Symbols enable a language user to distinguish "real" from "not real" and "present" from "not present." This separation between real and imaginary, present and absent, makes it possible for humans to separate "should do" from "does" and "should be" from "is." A parent may ask the youngest child, "Why can't you be more like your older brother?" The parent has an understanding made possible by the ability to organize experiences symbolically and to recognize that the younger child acts differently from the older child and from how the younger child *could* and *should* act. Humans are able to recognize the difference between what is, what was, and what should be.

The learning of symbols makes it possible for individuals to know how they are viewed from the standpoint of others. In time, most of us actually become objects of our own actions: We define and react to ourselves in some of the same ways that others have defined and reacted to us (Mead, 1934). We come to learn that we are sons or daughters, males or females, attractive or ugly, and overweight, skinny, or just right, and we treat ourselves accordingly. We view ourselves from the perspectives of others, and we take account of how they have treated us and develop a sense of what and who we are as a result.

Ideal Culture

Because humans are able to differentiate what is from what should be, they are able to formulate rules about "proper" and "improper" ways of acting, thinking, and feeling. These rules are called *norms*. The directive to "chew your food with the lips closed" is the statement of a norm, as is the sign found in many restaurants informing customers that "no checks are allowed." Norms are often coupled with *sanctions*. A positive sanction is a symbol that conveys approval and encourages norm-following activities; a negative sanction conveys disapproval and is designed to discourage normative violations. Norms are an important part of culture, and they help us to understand regularities in human behavior better. No important and enduring human activity is likely to be left entirely to chance; efforts will be made to organize it socially through rules, regulations, or informal understandings.

Some human activities can exist *somewhat* independently from a socially constructed system of rules. Eating, breathing, drinking, sleeping, and waste elimination are regulated in all places and in all times by rules and understandings, but these activities would *exist* even without rules. Rules that cover these activities are best described as regulative (Searle, 1995, p. 27). However, some kinds of human activity are *constituted* by rules, not merely regulated by them. If the rules were nonexistent, so would the activity in question be nonexistent. For example, if you refuse to follow the rules of chess, then you are not playing the game (Searle, 1995, pp. 27-29). Though deviance may reflect certain biological and/or psychological characteristics, it seems far more profitable to explore the many ways that social relationships, interpersonal reactions, and cultural understandings actually constitute the multiplicity of forms that deviance takes.

Certain groups, because of their luck, skill, determination, or control of important resources such as power or money, usually manage to influence disproportionately the content of culture by creating and spreading *ideologies* throughout a society (Bourdieu, 1991). An ideology is a set of ideas or beliefs that serves the interests of one segment of a society more than all other segments; an ideology legitimates some specific social arrangement. Many belief systems—racism, sexism, communism, socialism, nationalism, democratism, capitalism, Protestantism, Catholicism, Judaism—may be viewed as ideologies because they benefit one sexual group, economic group, political group, geographical group, social group, or religious group more than others. Ideological beliefs may be very attractive to some people and easy for them to accept (Boudon, 1994, p. xii). Only when conflicts erupt that challenge the legitimacy of groups with a monopoly of power do ideas that were once accepted as self-evident truths come to be seen as the self-serving rationalizations that they really are (Berger, 1995).

Most people in a society are familiar with its ideal culture; they may sometimes try to follow it themselves, and they may sometimes encourage others to follow it. People who do not follow the rules of ideal culture may be negatively sanctioned by others, or they may define themselves as less worthy because they have broken the rules and may experience guilt or embarrassment as a result (Scheff, 1990). Ideal culture gives the appearance of great uniformity and consistency within a culture; ideal culture may even give the false impression that only one proper way exists for all people to act, think, and feel. Ideal culture may become so important to individuals in a group that they condemn anybody who fails to live up to its standards.

It is possible for the ideal norms of one society or group to be drastically different from the ideal norms of another society or group. This means that even if people are acting properly with regard to one set of standards, from the standpoint of other people in different places (or from the standpoint of people who have different understandings), they are acting improperly. During the New Orleans Mardi Gras, certain forms of "creative deviance" (Douglas, Rasmussen, & Flanagan, 1977, p. 238) that might normally be forbidden become customary. One of these is "parade stripping," in which women expose their naked breasts to people on parade floats so that they will be thrown glass beads and trinkets. People who are not so accepting of the practice refer to these women as "beadwhores," and they tend to view parade stripping as unbridled exhibitionism. To those individuals who are caught up in the playful atmosphere of Mardi Gras, however, things are different (Forsyth, 1992, p. 395). Parade stripping represents a ritualized exchange of things of value. The float rider gets to see naked breasts, and the woman receives beads, trinkets, and confirmation that her breasts are grand enough to warrant a bestowing of gifts (Shrum & Kilburn, 1996, p. 444).

The standards of ideal culture are usually too lofty to be followed by most of the people most of the time, so all cultures contain standards for acting, feeling, and thinking that are more realistic and easier to achieve. These "real" rules encourage or allow people to do what is smart, practical, effective, fun, or easy, rather than what is proper (Freilich, 1991). For example, instead of following the Golden Rule, some people prefer to "look out for number one." Sometimes people will ignore cultural rules or the expectations or wishes of others and do only what is personally beneficial. In this case, they are doing what they like without reference to rules or to the interests of other people at all. Ideal culture, rather than being an important influence on human behavior, may simply be a system of standards to which members of a culture can refer in order to convince themselves that they regularly act in exemplary ways.

The Social Construction of Reality

Humans construct so much of the reality within which they live that it is difficult to identify one reality that exists for all people (Berger & Luckmann, 1966, p. 86; Sarbin & Kitsuse, 1994, p. 8). Every social world is complex, and it changes all the time in uncontrollable and unpredictable ways (Gove, 1994, pp. 365-366);

this fact means that every social world has a provisional quality (Troyer, 1992, p. 36). However, because social reality is so extensive and enveloping, it attains the status of a force of nature, and its conventional nature is easy enough to miss (Searle, 1995, p. 4). Because particular people with particular interests and resources come to imbue certain ways of doing things and certain preferences with great significance, certain social forms are justified as better than all others. It is usually impossible to know if this is true or if these claims are merely the ideological rumblings of some particular group.

Because most social encounters that people have are typical and ordinary, they do not call into question the taken-for-granted nature of human experience, and the apparent objectivity of social reality is continually reaffirmed in everyday interaction (Berger & Luckmann, 1966, p. 23). The friends that we choose and the relatives that we have are sufficiently like us that contacts with them do not present enough contradictory input to show us how flimsy conceptions of reality really are. Even the times that people do encounter reality-shattering experiences are too infrequent to refute their faith in the existence of an objective and concrete reality. Particular interpretations of human experiences can become so routinized that they seem to be self-evident and beyond dispute, and so, in this sense, they are.

Deviants are cast in such a way that their deviance really offers no serious threat to the dominant construction of reality. Deviance is branded as dysfunctional, unhealthy, evil, dangerous, or abnormal. The successful definition of alternate realities as inauthentic, pathological, or deviant reinforces the dominant view of reality and makes it appear more immutable and concrete than it actually is (Berger & Luckmann, 1966, pp. 112-115). The construction of deviant labels and their assignment to particular individuals serve to mask social conflicts and make the labelers more confident that their way is the *only* way (Parsons, 1951, p. 266).

Theoretical Views: The Old and the Not-So-Old

One of the earliest conceptions (which is still around) was founded on the belief that deviants could be separated from nondeviants and deviance from nondeviance on the basis of inherent or intrinsic characteristics (Gibbs, 1966). Usually, some biological, psychological, or sociological factor such as body chemistry, intelligence, or social disorganization was identified and then blamed for the existence of the deviance (Vold, Bernard, & Snipes, 1998). Because the deviance was almost always viewed as unacceptable and unnecessary, the temptation to see defect, abnormality, or degeneracy in the biology, psychology, or social situation of deviants was too great to resist. Sometimes the acts themselves were classified as inherently or intrinsically deviant. Parsons (1951, p. 250) insisted that deviance could be identified by its potential to produce a disturbance in the equilibrium of interactive systems, and Schwendinger and Schwendinger (1975) argued that deviance is behavior that harms individuals.

But as more and more was learned about deviance, and as sociologists and anthropologists got more involved in its study and demonstrated the great variety of human customs and experiences, the meaning of deviance changed. New conceptions were developed that were not predicated on a belief in the existence of intrinsic characteristics of either deviance or deviants. Sumner's (1906) discussion of mores, folkways, and other social rules declared that they are inherited from the past and that they direct human behavior almost automatically (pp. 76-77). When they are followed, they facilitate the adjustment of individuals to life conditions and to the particular demands of the time and the place (p. 58). It was an easy leap to the view that deviance is a normative departure (Gibbs, 1966, p. 14).

Sumner (1906, pp. 521-522) believed that social rules could make anything right and prevent the condemnation of anything. If this is true, norms could also make anything wrong. According to Becker (1963), groups create deviance by making rules whose violation qualifies as deviance, identifying rule breakers, and treating them as outsiders (p. 9).

> We must see deviance, and the outsiders who personify the abstract conception, as a consequence of a process of interaction between people, some of whom in the service of their own interests make and enforce rules which catch others who, in the service of their own interests, have committed acts which are labeled deviant. (p. 163)

The creation and enforcement of rules is a moral enterprise that depends on the willingness of some individuals to go to the time and trouble to get their particular view of right and wrong adopted by others.

> A successful, and enforceable, social construction of a particular label of deviance depends on the ability of one, or more, groups to use (or generate) enough power so as to enforce *their* definition and version of morality on others . . . Deviance . . . always results from negotiations about morality *and* the configuration of power relationships. (Ben-Yehuda, 1990, pp. 6-7)

Some crusades may be very successful and have enduring effects; other crusades may be short-lived, dying quickly and with little fanfare (Becker, 1963, pp. 147-163).

The meaning of *norm* changed. No longer were norms necessarily viewed as reflective of a societywide consensus, as embodying a shared morality, or even as essential designs for living. C. Wright Mills (1943, p. 170) showed us that the prevailing norms almost always reflect some specific group's biased view of what is proper and what is improper; norms reflect the power, the interests, and the outlooks of the groups that create them. The greater the social conflict and cultural heterogeneity, the less likely it is that any normative system could even come close to reflecting a universal consensus or agreement on the proper and improper ways of acting, thinking, feeling, and being.

Once theoretical explanations of deviance evolved to the point where they were sensitive to the role played by the "other" in the construction of deviance, a whole new world of possibilities was opened up. It could be maintained with credibility and authority that social control itself has the ironic effect of actually creating deviance and channeling the direction that it takes (Lemert, 1972, p. ix). Tannenbaum (1938) insisted that social labels and other social reactions actually create deviance: "The process of making the criminal, therefore, is a process of tagging, defining, identifying, segregating, describing, emphasizing, making conscious and self-conscious; it becomes a way of stimulating, suggesting, emphasizing, and evoking the very traits that are complained of" (pp. 19-20). A new wrinkle had been added: The understanding of social deviance required an analysis of the processes by which persons came to be defined and treated as deviant by others. The definition of deviance changed to reflect this new under-standing: "Deviance is not a property *inherent in* certain forms of behavior; it is a property *conferred upon* these forms by the audiences which directly or indirectly witness them" (Erikson, 1962, p. 308). Characterizations of behaviors and attributes became the principal target of study, not the behaviors and attributes themselves (Schur, 1975, p. 287).

The assertion that social control can create deviance can mean many things; however, two possibilities stand out. First, social control may identify something as deviant and separate it from other behaviors or attributes that are not considered deviant. Something so identified may be called "deviancy by definition."

> Deviance may be conceived as a process by which the members of a group, community, or society (1) interpret behavior as deviant, (2) define persons who so behave as a certain kind of deviant, and (3) accord them the treatment considered appropriate to such deviants. (Kitsuse, 1962, p. 248)

A second possibility is that the reactions of some people—what they say, do, or believe—can propel other people in the direction of greater involvement with deviant pursuits. Such involvement may be called "secondary deviance."

> [W]e start with the idea that persons and groups are differentiated in various ways, some of which result in social penalties, rejection and segregation. These penalties and segregative reactions of society or the community are dynamic factors which increase, decrease, and condition the form which the initial differentiation or deviation takes. (Lemert, 1951, p. 22)

According to the secondary deviance proposition, an individual's deviance can be channeled or even amplified by the reactions of others if the social dynamics are right, but this is not the same thing as constructing deviancy through naming, classifying, and judging some behaviors and attributes as proper and others as improper. The secondary deviance proposition is about the ironic nature of social control: Social censure can actually cause more deviance, not less. Deviancy by definition is about the arbitrary nature of social control: Social censure is the event that transforms human diversity into social deviance by altering how behaviors and attributes are defined and perceived (Sumner, 1994, p. 222).

Deviance came to be viewed as an inevitable and rather ordinary feature of life in a pluralistic society, and deviants came to be viewed as more sinned against than sinning. Empathizing with the deviant who had been labeled, stigmatized, and forced to associate with other deviants became a very legitimate enterprise in the sociology of deviance (Becker, 1967). In fact, siding with deviants became as defensible as siding with representatives of conventional society such as police, judges, or psychiatrists, and the deviant's right to be different and to be free from stigma and harassment was actively defended. Deviance came to be viewed in political terms, and power—the power to label and the power to legitimate one's own view of proper and improper in some hierarchy of credibility (Becker, 1967)—was identified as *the* critical resource that allowed some people to benefit themselves by transforming the actions and attributes of others into something strange or even despicable.

Whose side *are* we on? The emergence of a radical view of deviance meant that some theorists took the side of the deviants with a vengeance. Not only did these theorists defend the right of deviants to be different, but they condemned representatives of conventional society and branded them as the dangerous, odd, or misguided ones. Life in an unequal, competitive, insecure society, they believed, was brutalizing for some people, and brutal conditions generate brutal behaviors. Radicals viewed deviance as one of the choices that people consciously make as one possible solution to the difficulties posed for them by life in a contradictory society (Taylor, Walton, & Young, 1973, p. 271). Radicals insisted that the control of deviants and the suppression of deviance are both principal ways that threats to the economic and political systems are counteracted and that the status quo is preserved for the benefit of powerful groups (Quinney, 1974, p. 52). Quinney (1973, p. 60) went so far as to claim that the really bad people are those who make laws to protect their own interests and to legitimate the repressive social control of powerless groups.

Deviance is a construction of social actors, and social meanings of deviance are always problematic, just like the social world itself (Lyman & Scott, 1989, p. 7). We must always have a clear understanding of exactly which actors are deciding that something is deviant, why they are making the claims that they are (Spector & Kitsuse, 1977), and their level of success in convincing others that they are really authorities on the issue in question (Best, 1990, pp. 11-13). Each relationship contains rules, understandings, and background assumptions that might very well seem odd to people from different relationships (Denzin, 1970, pp. 131-132).

Relativity and Social Deviance

What Is Relativity?

Relativity is first and foremost a method for understanding human behavior. It exists when a visitor to some group makes every effort to understand its members' customs and outlooks from their viewpoint. Relativity requires the ability to suspend one's egocentrism and ethnocentrism sufficiently that one's vision is clear and unclouded enough to enable one to know what is happening and describe it from the standpoint of the people experiencing it. We must be familiar with actual social relationships in the social worlds in which they occur (Douglas & Waksler, 1982, pp. 22-25). This may require an observer to overcome language barriers and great differences in outlooks and understandings.

A relativist is less likely to be ethnocentric and egocentric because he or she usually understands that what is acceptable in one place and time is not acceptable in another place and time and that evaluations of human behavior vary dramatically from situation to situation. Goffman (1961) indicated how this can work, stating that "the awesomeness, distastefulness, and barbarity of a foreign culture can decrease to the degree that the student becomes familiar with the point of view to life that is taken by his [or her] subjects" (p. 130). Relativists understand that the standards and understandings of one time and place cannot be transferred indiscriminately to another time and place without producing a great deal of misunderstanding and distortion.

A competing view to relativism goes by the name of the *objectivist* or *absolutist* approach (Simon, 1996, pp. 288-290). Objectivists or absolutists identify norms or rights that they believe are good, necessary, or healthy for all people in all places and at all times. Anything that runs afoul of these norms or violates these rights is called deviance, and the people responsible for the trespass are automatically deviant. Social reactions—whether anyone knows that rules have been broken or even whether anyone cares—are considered to be unimportant in deciding who or what is deviant. A variant of the objectivist approach posits the existence of invariant traits, almost always viewed as personal defects, to explain why some people do things to hurt or upset others: deviance is what people with inherent abnormalities do.

It would be a mistake to exaggerate the role of either biological or natural laws in variations in human social behavior and in evaluations of these patterns. Biology and the physical world certainly must be reckoned with by humans (Udry, 1995, p. 1269; Walsh, 1995, pp. 173-199), but human experiences and our social relationships are not reducible to the simple operation of biological or natural laws (Rothman, 1995). Our connectedness with each other is a human construction—a reflection of cultural learning and socialization experiences—not some behavior pattern that we have inherited from our nonhuman ancestors (Maryanski & Turner, 1992, p. 163). If any human universals have relevance for our understanding of social deviance, then they surely must be the human predilections to identify and/or construct social differences, to evaluate those differences, and then to persecute those individuals who seem to be out of step with everybody else (Moore, 1987).

In Defense of Relativity

Adopting the approach of relativity does not mean that every practice, no matter what it is or when or where it occurs, has to be applauded. Relativity does not require moral indifference, and it does not mean that one can never be upset or horrified by what one experiences in another group or culture (Goode & Ben-Yehuda, 1994, p. 73). Relativity just reminds us that our personal beliefs or our cultural understandings are not necessarily found everywhere. If we condemn some practice, we must be certain that we are doing it for reasons other than our own particular ethnocentric or egocentric biases. Our biases cannot be avoided completely, but relativity allows us the best chance to recognize them for the partialities that they are.

Relativity could become obsessive (and therefore irresponsible) and lead to a blind acceptance and tolerance of what other people in some group are doing, no matter what it is. Not only must we avoid ethnocentrism and egocentrism, but we must avoid romanticizing diversity and viewing all differences as somehow equivalent and valid. As Matza (1969) suggested, a sow's ear is not the same thing as a silk purse, no matter how much we would like it to be (p. 44). Blind acceptance of cultures other than our own would actually hamper our ability to understand social deviance. Social scientists are required by the demands of their disciplines to understand as fully as possible the social systems that they are studying. This means that they will almost always have to uncover and critically evaluate pretensions and propaganda people use to hide, distort, or legitimate what they are doing (Berger, 1963, p. 38). Carried to extremes, romanticizing or glorifying diversity will have the unfortunate outcome of making it more difficult to see and deal with the extreme suffering that deviance can cause in the lives of people who are adversely affected by it (Matza, 1969, p. 44).

The Serious Implications of
Taking Relativity Seriously

Change is the only real constant, and social life is transitory, ambiguous, and conflictual (Gibbs, 1994). Whatever exists could be otherwise, and no matter how something is evaluated, it can always be evaluated differently. Practically anything that can be done with the human body has been done with the human body and as a source of pleasure to somebody, somewhere, sometime; and practically any claim that can be made about what people have done with their bodies has been made. Humans have a remarkable capacity to make practically anything proper or improper, as the case may be. The world is constantly in flux, and people have done, are doing, and will continue to do a multitude of things that will lead to the delight, indifference, or dismay of others.

A colleague of mine once observed that the ties that bind—traditions and customs, ways of doing things, standard notions of proper and improper—also qualify as the ties that *blind*. Human beings assign meaning to practically everything that can be seen, felt, smelled, tasted, or heard. Once this is done, it tends to blind them to other possibilities and to alternate ways of thinking, feeling, or acting. One individual's object of outrage or disgust might be, and probably is, another individual's object of reverence, worship, or desire; one individual's revulsion might be another individual's attraction; and the object or objects of hatred for members of one human society just might be, and probably are (or have been), the objects of adoration and affection for the members of another society (P. Winther, personal communication, 1994).

Culture shock is the feeling of disorientation or uneasiness that one might experience when traveling to a different place and being forced to accommodate to new ways of thinking, feeling, or acting. Sometimes the culture shock is mild. If you went from a hot climate, such as one finds in California or Florida, to a much colder climate, such as one finds in Illinois or New York, you might experience things in the winter to which you were unaccustomed. People would seem to be dressing in unusual ways and doing unusual things such as buying snow blowers and snow shovels, but it would be easy to adjust to these changes, and they would be only temporarily disorienting. What if you were told, however, that the stew you had just eaten was made with dog meat, horse meat, sheep eyes, the flesh of recently killed rattlesnakes, and a fruit bat? How would you feel if you were told that the crunchy snack you had just eaten was not a party mix but baked locusts and termites? In these cases, culture shock may be acute and quite unsettling. Sometimes it is an experience an individual will never forget.

Deviance has the potential to be shocking. As we travel to other places in other lands, or to different places within our own land, we will always find different customs. This means that if we journey to many different places, we will have a greater chance of experiencing something shocking or unsettling. This culture shock may propel us in the direction of greater ethnocentrism and egocentricity. No way, we tell ourselves, would we ever do *that*; we would have to be sick, crazy, or out of our heads. Yet we must fight this centeredness and make every effort to understand the context of action. We must understand that if we had been raised in a culture where people eat dogs, horses, sheep eyes, rattlesnakes, locusts, termites, or fruit bats, we would, in all likelihood, be eating and enjoying them too.

Viewing deviance as something that can produce culture shock but that is not intrinsically bad or sick is extremely valuable for us. Just as we might study any form of behavior in a foreign land, we can study deviance "at home," even in our own homes, as part of changing social relationships. Our goal is to understand the context of behavior and how and why behavior is defined and evaluated the way it is. We will not spend time trying to decide if deviance is abnormal or sick and in need or cure or correction. These are important questions for some people but not for us. Just as it would be too ethnocentric to conclude that eating dog meat instead of cow meat is abnormal or sick, it is too ethnocentric to conclude that people who inhale certain kinds of substances (e.g., marijuana) are more abnormal or sicker than people who inhale other kinds of substances (e.g., tobacco). Deviance in practically all its forms is a normal feature of human societies, and we must fight the temptation to equate deviance and diversity with disease or abnormality. Some deviance is unsettling or shocking to people who are unfamiliar with it, but this does not mean that deviance is necessarily abnormal or sick everywhere and at all times.

Summary and Conclusions

The egocentricity of childhood, in which we are concerned principally with ourselves, is eventually replaced by the ethnocentricity of adulthood, in which we come to evaluate the customs and standards of other people against those of our own groups. Cultural standards and ethnocentrism are acquired in the context of socialization, in which each individual learns the social heritage of his or her society. An important part of this process is the acquisition of symbols and the language of which they are a part. The great diversity in human behavior is produced by the interplay of biological and social factors.

Every society contains an ideal culture—shared understandings about exemplary ways of acting, thinking, and feeling. Individuals in a society also understand that they can sometimes depart from the ideal culture without being punished or negatively sanctioned. Humans create social forms as they act together, and the varied nature of human experience means that no one universal and uniform reality exists.

Theories of deviance have evolved over time. Early views were based on a belief that intrinsic characteristics separated both deviants and deviance from their opposites. Attention ultimately shifted to norms and then to labels and social reactions. Social deviance came to be viewed as a regular feature of life in a pluralistic society that could be caused by social control itself. Deviance is a social construction that emerges from social differentiation, social conflict, and social disagreement. The meanings of deviance are always problematic.

A great deal of diversity can be found as we move from society to society or group to group. Relativity is a way of examining standards and customs by understanding their context. Deviance, like beauty, is in the eye of the beholder, and it exists because some groups decide that other groups ought not to be doing what they are. Deviance results from dynamic relationships among many people; it is not an unchanging or immutable condition with intrinsic or inherent qualities. We must remember that all things are transitory and impermanent, including human understandings about proper and improper ways of acting, thinking, and feeling.

References

Becker, H. (1963). *Outsiders: Studies in the sociology of deviance.* New York: Free Press.

Becker, H. (1967). Whose side are we on? *Social Problems, 14,* 239-247.

Ben-Yehuda, N. (1990). *The politics and morality of deviance: Moral panics, drug abuse, deviant science, and reversed stigmatization.* Albany: State University of New York Press.

Berger, B. (1995). *An essay on culture: Symbolic structure and social structure.* Berkeley: University of California Press.

Berger, P. (1963). *Invitation to sociology: A humanistic perspective.* Garden City, NY: Anchor.

Berger, P. L., & Luckmann, T. (1966). *The social construction of reality.* Garden City, NY: Anchor.

Best, J. (1990). *Threatened children: Rhetoric and concern about child victims.* Chicago: University of Chicago Press.

Boudon, R. (1994). *The art of self-expression: The social explanation of false beliefs* (M. Slater, Trans.). Cambridge, UK: Polity.

Bourdieu, P. (1991). *Language and symbolic power* (G. Raymond & M. Adamson, Trans.). Cambridge, MA: Harvard University Press.

Denzin, N. (1970). Rules of conduct and the study of deviant behavior: Some notes on the social relationship. In J. Douglas (Ed.), *Deviance and respectability: The social construction of moral meanings* (pp. 120-159). New York: Basic Books.

Douglas, J., Rasmussen, P., & Flanagan, C. (1977). *The nude beach.* Beverly Hills, CA: Sage.

Douglas, J., & Waksler, F. (1982). *The sociology of deviance.* Boston: Little, Brown.

Erikson, K. (1962). Notes on the sociology of deviance. *Social Problems, 9,* 307-314.

Forsyth, C. (1992). Parade strippers: A note on being naked in public. *Deviant Behavior, 13,* 391-403.

Freilich, M. (1991). Smart rules and proper rules: A journey through deviance. In M. Freilich, D. Raybeck, & J. Savishinsky (Eds.), *Deviance: Anthropological perspectives* (pp. 27-50). New York: Bergin & Garvey.

Gibbs, J. (1966). Conceptions of deviant behavior: The old and the new. *Pacific Sociological Review, 9,* 9-14.

Gibbs, J. (1994). *A theory about control.* Boulder, CO: Westview.

Goffman, E. (1961). *Asylums.* Garden City, NY: Doubleday.

Goode, E., & Ben-Yehuda, N. (1994). *Moral panics: The social construction of deviance.* Oxford, UK: Basil Blackwell.

Gove, W. (1994). Why we do what we do: A biopsychosocial theory of human motivation. *Social Forces, 73,* 363-394.

Kitsuse, J. (1962). Societal reaction to deviant behavior: Problems of theory and method. *Social Problems, 9,* 247-257.

Kluckhohn, C. (1949). *Mirror for man: The relation of anthropology to modern life.* New York: McGraw-Hill.

Kumbasar, E., Romney, A. K., & Batchelder, W. H. (1994). Systematic biases in social perception. *American Journal of Sociology, 100,* 477-505.

Lemert, E. (1951). *Social pathology.* New York: McGraw-Hill.

Lemert, E. (1972). *Human deviance, social problems, and social control* (2nd ed.). Englewood Cliffs, NJ: Prentice Hall.

Lyman, S., & Scott, M. (1989). *A sociology of the absurd* (2nd ed.). New York: General Hall.

Margolin, L. (1993). Goodness personified: The emergence of gifted children. *Social Problems, 40,* 510-532.

Maryanski, A., & Turner, J. (1992). *The social cage: Human nature and the evolution of society.* Stanford, CA: Stanford University Press.

Matza, D. (1969). *Becoming deviant.* Englewood Cliffs, NJ: Prentice Hall.

Mead, G. H. (1934). *Mind, self, and society* (C. W. Morris, Ed.). Chicago: University of Chicago Press.

Mills, C. W. (1943). The professional ideology of social pathologists. *American Journal of Sociology, 49,* 165-180.

Moore, R. I. (1987). *The formation of a persecuting society: Power and deviance in western Europe, 950-1250.* New York: Basil Blackwell.

Parsons, T. (1951). *The social system.* New York: Free Press.

Piaget, J. (1948). *The moral judgment of the child.* New York: Free Press.

Quinney, R. (1973). There are a lot of folks grateful to the Lone Ranger: With some notes on the rise and fall of American criminology. *Insurgent Sociologist, 4,* 56-64.

Quinney, R. (1974). *Critique of the legal order: Crime control in capitalist society.* Boston: Little, Brown.

Rothman, B. K. (1995). Of maps and imaginations: Sociology confronts the genome. *Social Problems, 42,* 1-10.

Sarbin, T., & Kitsuse, J. (1994). Prologue. In T. Sarbin & J. Kitsuse (Eds.), *Constructing the social.* Thousand Oaks, CA: Sage.

Scheff, T. (1990). *Microsociology: Discourse, emotion, and social structure.* Chicago: University of Chicago Press.

Schur, E. (1971). *Labeling deviant behavior: Its sociological implications.* New York: Harper & Row.

Schur, E. (1975). Comments. In W. Gove (Ed.), *The labelling of deviance: Evaluating a perspective* (pp. 285-294). Beverly Hills, CA: Sage.

Schwendinger, H., & Schwendinger, J. (1975). Defenders of order or guardians of human rights? In I. Taylor, P. Walton, & J. Young (Eds.), *Critical criminology* (pp. 113-146). London: Routledge & Kegan Paul.

Searle, J. (1995). *The construction of social reality.* New York: Free Press.

Sheley, J., & Corsino, L. (1994). Tunnel-visioned activity and sociologically problematic deviant behavior. *Deviant Behavior, 15,* 267-288.

Shrum, W., & Kilburn, J. (1996). Ritual disrobement at Mardi Gras: Ceremonial exchange and moral order. *Social Forces, 72,* 423-458.

Simon, D. (1996). *Elite deviance* (5th ed.). Boston: Allyn & Bacon.

Spector, M., & Kitsuse, J. (1977). *Constructing social problems.* Menlo Park, CA: Cummings.

Sumner, C. (1994). *The sociology of deviance: An obituary.* New York: Continuum.

Sumner, W. G. (1906). *Folkways: A study of the sociological importance of usages, manners, customs, mores, and morals.* Boston: Ginn.

Tannenbaum, F. (1938). *Crime and the community.* Boston: Ginn.

Taylor, I., Walton, P., & Young, J. (1973). *The new criminology: For a social theory of deviance.* New York: Harper & Row.

Troyer, R. (1992). Some consequences of contextual constructionism. *Social Problems, 39,* 35-37.

Udry, J. R. (1995). Sociology and biology: What biology do sociologists need to know? *Social Forces, 73,* 1267-1278.

Vold, G., Bernard, T. J., & Snipes, J. (1998). *Theoretical criminology* (4th ed.). New York: Oxford University Press.

Walsh, A. (1995). *Biosociology: An emerging paradigm.* New York: Praeger.

Zerubavel, E. (1991). *The fine line: Making distinctions in everyday life.* New York: Free Press.

2

Being Deviant

Introduction: Status and Stigma

Over six generations ago, Martin Fugate and his bride settled on the banks of eastern Kentucky's Troublesome Creek. They had children, who had children, who had children. Most of them were healthy and lived well into old age. Not terribly long ago, Martin Fugate's great-great-great-great grandson was born. The boy was as healthy as a newborn could be, but he did have one curious trait: dark blue skin, the color of a plum or denim blue jeans. The attending physicians were concerned. Did the child have a blood disorder? The child's grandmother told them not to worry; many of the Fugates had blue skin (Trost, 1982).

A young hematologist from the University of Kentucky was curious about the reason for the unusual skin color. With the help of a nurse (who shared his interest in the blue skin) and a blue couple, the physician eventually uncovered the reason for the blue tint. It was caused by a hereditary condition that allowed too much methemoglobin (which is blue) to accumulate in the blood. The blue people, it seemed, lacked an enzyme that is necessary for the regulation of methemoglobin. It was either a quirk of fate or an affair of the heart that accounted for the original spark that generated the blueness of the Fugates: Martin Fugate married a woman who carried the same recessive gene for blueness that he did. Because members

21

of the Fugate line were content to remain where they were born, people with the recessive gene that caused the blueness often married and had children with others who had the same hereditary trait. As a result, the number of blue people in this region of Kentucky increased.

The physician and the nurse had more than an academic interest in the blue skin color. As befitted their medical training, they really wanted to find a "cure" for it. Once they knew the reason for the skin color, it was easy enough for them to find an "antidote." Methylene blue is a chemical that changes the color of methemoglobin. When it was injected into the blue people, it had the desired effect: Their skin turned pink. However, for the transformation to last, the former blue people of Kentucky would have to take a pill every day because the effects of methylene blue are short-lived (Curra, 1994, pp. 19-21).

People can be and often are labeled as deviants for attributes that they possess, not just what they do. The experiences of people who are treated as deviant for what they are—their shame, self-loathing, and social isolation—may not be appreciably different from the experiences of people who are treated as deviant for what they do. Both human behavior and human beings must be part of a definition of deviance because deviance is sometimes a matter of being, or being and doing, rather than exclusively doing (Sagarin, 1975, p. 9). The recognition that people are labeled and negatively sanctioned for what they are is valuable (Bowditch, 1993, p. 500). It fleshes out our understanding of the full scope of the deviant experience. What could show the situational and variable nature of deviance better than the fact the people are stigmatized for characteristics over which they have little or no control? If being blue among a bunch of other blue people can be labeled as deviant and in need of change, anything can be.

Personal Appearance and the Sociocultural Matrix

Physical appearance is one of the more visual and immediate cues that people bring to encounters with others. It is used as a signifier of other, more difficult-to-measure personal factors, and it plays an important role in patterns of social interaction and in the differential treatment that people receive. The determination that someone is beautiful or ugly (or any gradation in between) involves selective viewing, definition, classification, and evaluation. Beauty, just like deviance, may be in the eyes of the beholder.

Studies have been done in which images of composite faces were constructed on a computer from individual pictures of faces. The greater the number of pictures of faces that went into the construction of the composite photo, the more average or symmetrical the composite became. So, if eight separate photos were used to make the composite, the composite was more symmetrical, typical, or average than if four separate photos had been used to make it. The central finding was that the composite image constructed from 32 separate pictures of faces was ranked as more attractive than a composite made from 16, 8, or 4 separate pictures of faces (Buss, 1994, p. 54).

The equation of symmetry with beauty and asymmetry with ugliness—what we can call the "damaged-goods" theory of attractiveness and mate selection—is interesting but flawed. People use things other than the face to determine attractiveness (figure, health, character, personality, age), and we can certainly think of times when a symmetrical face would *not* be judged particularly attractive (Jabba the Hut seemed symmetrical to me, as did Darth Vader, but I wouldn't want my sister to marry either of them). It is not entirely clear what symmetry actually represents. It seems to be shorthand for average or normal: The greater the number of individual photos used to construct the composite, the more typical or average it becomes. Yet most of us suspect that the world's "most beautiful people" have gotten that way by more than being average. An over-arching tautology seems to be operating that beautiful people are beautiful because they are so darn attractive. Maybe it's true, but it's not very helpful.

Using *symmetrical* to mean "average" is not the same as using *symmetrical* to mean "ideal" or "flawless," and perhaps "ideal" or "flawless" is closer to what is meant by *beautiful*. When American men and women are given the task of constructing an ideal or perfect human face, they do seem to agree. When asked to construct an image of the perfect female face, they come up with a very youthful face with full lips and a narrow mouth (Small, 1995, p. 143). The perfect male face (according to American female undergraduates) has large eyes, a large chin, a small nose, and prominent cheekbones. It is an open, pleasant face with rugged features (Small, 1995, p. 143). None of these traits seems extraordinary, but maybe it is unusual to find so many diverse indicators of female attractiveness on the same female or of male attractiveness on the same male.

Are uniform and universal cues for beauty and ugliness to be found? Some traits—festering sores, hacking coughs, incessant sneezes, unpleasant body odors—are probably widely defined as unattractive or ugly, and they could easily repel potential mates in practically any situation. Other traits—health, youth, vitality, sincerity, integrity, kindness, poise, intelligence, joviality—are likely to be viewed as attractive, and they may be used to classify people as beautiful. However, that is a far cry from the claim made by some that objective and universal beauty cues exist that have been programmed into our biology and psychology to ensure reproductive success (Buss, 1994, pp. 53-54; Ridley, 1993, p. 280). Just because people say that they find some particular qualities attractive in a mate does not necessarily mean that their fantasies are echoes of some genetic predisposition established long ago (Small, 1995, p. 148).

How beautiful or ugly a partner looks depends on our needs, interests, and the nature of our relationship to him or her and, importantly, our relationship to other "hims" or "hers." We may have an ideal image of beauty in our minds, but it is usually broad and alterable. Though age and health are supposed to be important factors in long-term relationships oriented toward reproduction, it may be just those relationships in which they are the least important. People do not usually divorce or separate on the grounds that their partner is too old or too sick. If anything, growing old together may enhance the quality of the relationship for both partners.

It is almost certain that if we looked, we could find *both* beautiful and ugly things in every person on earth, and a trait that looks ugly at one time may look beautiful at another. We might even be drawn to certain traits of an individual because we are repelled by other traits of that *same* individual. Learning must certainly play a big role in what and whom people find attractive, and culture must have an impact on what a group defines as beautiful or ugly that parallels or surpasses any influence of biology. We want mates who are enjoyable to be around, who will make us feel special and needed, and who are accommodating enough and sensitive enough to be responsive to our wants and needs. In the absence of these, a partner's youth, health, fecundity, or high social status will matter very little.

Human relationships involve a great deal of compromise and negotiation, and what we really want is not always what we really get. Most of us realize that our real partner will turn out to be different from our ideal partner. People tend to mate with people like themselves in terms of physical, psychological, geographical, social, and emotional characteristics. A strong correlation exists between how we view ourselves and what we think we can get in the search for possible mates (Small, 1995, p. 152). What draws us to a relationship may be very different from what keeps us together, and the evaluation of partners is a flexible and ongoing enterprise.

All the different body forms, skin colors, nose shapes, ear designs, and facial configurations, along with all the deliberate modifications of the human form, make it hard to believe that any universal and uniform standard of beauty and ugliness could ever be found. Certainly, it is not hard to find examples that challenge the claim that standards for beauty are universal and programmed into us at birth.

A cross-cultural survey of notions of beauty is sure to include such "oddities" as a preference for cross-eyes (Mayans), flattened heads (Kwakiutl), black gums and tongue (Maasai), black teeth (Yapese), joined eyebrows (Syrians), absence of eyebrows and eyelashes (Mongo), enormously protruding navels (Ila), pendulous breasts (Ganda), gigantic buttocks (Hottentot), fat calves (Tiv), crippled feet (Chinese), and so on. (Gregersen, 1983, p. 81)

The erotic potential of female genitalia is found throughout the world, but the Kgatla-Tswana (Africa) put their own spin on the whole affair:

With the onset of puberty Kgatla girls start pulling their labia and sometimes will ask a girlfriend to help. If the labia do not get longer as quickly as desired, the girls resort to magic. They kill a bat and cut off its wings, which they then burn. The ashes are ground up and mixed with fat. Each girl makes little cuts around her labia and smears the bat-ash ointment into the cuts. (Gregersen, 1983, p. 92)

This little bit of magic is designed to get the labia to grow quickly to the size of the wings of a bat.

Teeth are an important part of one's appearance, and having a "nice" smile is usually considered an asset. The existence of "best smile" contests suggests that judges know a great smile when they see one (or believe that they do). Yet what qualifies as a "great smile" varies across the globe. Teeth have been permanently colored, knocked out, dug out, filed down, decorated, drilled, and chipped in order to heighten their attractiveness (Gregersen, 1983, p. 97). U.S. models and actresses (and others, thanks to the influence of television advertising) spend time and money on the whitening of their teeth. The Nilotes of East Africa would find these efforts at whitening incomprehensible. Beauty for them involves knocking out the lower front teeth (up to six), usually at the start of adolescence (Gregersen, 1983, p. 97).

It is a long way from laboratory studies of beauty/symmetry to the real world of finding and uniting with beautiful people. A few things—a very few things—*might* be universally viewed as beautiful or as ugly. However, what is important is the total package a person brings to a relationship—or another's perception of that package. Having a beautiful face—or any other specific trait—is not that important, at least not for long, and at least not in long-term relationships. Many things—opportunity, accessibility, availability, personal objectives, individual motives, physical qualities, psychological and emotional factors, personality characteristics, social attributes such as status—are used by people as they form relationships. Humans do find a wide range of things attractive and beautiful, some of which reflect cultural meanings and some of which reflect more idiosyncratic preferences. No matter who we are with or what we find attractive, most of us realize that if we had come to a different fork in the road and taken a different path, things would have been different.

The Social Construction of Spoiled Identities

Identity and Social Stigma

Each of us is a cluster of different attributes—our identity—and each of us engages in a large number of social acts during the course of our lives. We must wonder, therefore, why certain designations are used more often than others to describe us, and why some of them stick more readily to us in describing what we are and what we do. Hughes (1945) coined the term *master status* to describe a status that evolves into the dominant way an individual is interpreted or classified by others. Sex is usually a master status, as are skin color and occupation. These are major identity pegs, and they play a role in most human relationships. The status of deviant, Becker (1963) informed us, is also a master status. If an individual is defined as a deviant, this status can predominate over many others that the individual occupies and become a controlling one in the eyes of the beholders (pp. 33-34). A strong probability exists that as a deviant identity evolves into a master status, the level of social censure will increase and the sanctioned individual will experience stigma.

What is stigma? At one time, the word *stigma* was used to identify a distinguishing mark or brand cut into an individual's flesh for the purpose of identifying him or her as a tainted or despised individual. Now *stigma* means *any* attribute—a physical sign or character clue—that is accompanied by shame or disgrace. Too-tall Jones is identified by body, and Otis the town drunk is identified by character. Stigma can also have little directly to do with body or character: It exists because a person is in the wrong place at the wrong time. A college professor might feel very comfortable in a college library on a Saturday night (and be defined by others as being in the right place at a reasonable time); a college student, however, might feel very uncomfortable in a library during recreational hours because it might suggest that he or she cannot find anywhere else to go. What a person is (or is perceived to be) can simply be out of synchronization with the situation, and this discrepancy is what causes all the fuss; in a different situation, all would be well. Rudolph the Red-Nosed Reindeer's discrediting nose became an object of reverence only because the situation changed.

Goffman (1963) reminded us that when speaking of stigma, what is really needed is a language of relationships rather than attributes because no attribute is automatically crediting or discrediting; stigma always involves a relationship between an attribute and its perception and symbolization by others (pp. 3-4). In other words, "normal" and "stigmatized" are not persons but perspectives on persons (p. 138). Whenever we discuss stigmatizing attributes, their relational nature must be kept clearly in mind.

Some of the social signals that people regularly give off are not regularly used by others as sources of information. The shape of an individual's nose is probably less important as an identity cue to most observers than information about a person's sex, occupation, or education. However, some social signals are important sources of social information (Goffman, 1963, pp. 43-44). Some signals are *prestige symbols*: They increase a person's standing in the estimation of others. Other signals (or the same signals in different situations) are *stigma symbols*: They serve to defame or devalue an individual. One other set of social signals is called *disidentifiers*, which serve to throw some doubt on an image one is projecting to others. Remembering the relational nature of attributes, we may still predict that in the United States a Rolex watch will be a prestige symbol; an arrest record will be a stigma symbol; and poor grammar for a college professor or refined speech for the village idiot will be a disidentifier.

The social construction of a spoiled identity can include a process that is called *retrospective interpretation.* In retrospective interpretation, an accused deviant's personal biography is scrutinized, and what is learned is used to reinterpret the identity of the individual and his or her situation (Kitsuse, 1962, p. 253). The rule breaker is recast in the eyes of others (Schur, 1971, p. 52), and he or she becomes a brand-new kind of person. The change in identities involves the destruction of one social self and its replacement by a totally different one (Garfinkel, 1956). What was once viewed as normal in the identity of the individual comes to be viewed as a facade or charade that was actually hiding a deeper, more concrete, and a more sinister constellation of traits.

The process of retrospective interpretation may be one of discovery. Why is he or she this way? If a kernel of truth is found in the accused deviant's personal history, it becomes part of the reconstructed biography of a new deviant. We may discover, for example, that a mass murderer has had an abusive childhood and likes guns very much, information that would never have come to light—or have been given much significance—if the murderer had not committed random acts of senseless violence. This process of discovery, to be sure, can be very selective, and factors that might refute, challenge, or complicate the biographical sketch may be ignored, dismissed, or downgraded. It is possible, however, for the process of biographical reconstruction to involve something else: invention or fabrication. For example, a deviant's parents might be perceived to have been abusive when they really were not.

People observe themselves, and they usually have their own opinions and make their own evaluations of their personal attributes. Just because other people may condemn an individual does not mean that he or she must share their opinions. Individuals may take pleasure and pride in what they are precisely because some other people do condemn it, or they may simply be indifferent to the reactions of others and march to the beat of a different drummer. People with attributes that are devalued by others can maintain a positive self-image not only by hiding or covering the troubling attribute but by believing that their personal attribute is actually a good and valued trait, regardless of what others may think (Kitsuse, 1980, p. 8).

At one time, people with certain spoiled identities—they used to be called "freaks"—could find a place in the world of popular entertainment and amuse- ment (Bogdan, 1988). Siamese twins, bearded ladies, tattooed men, giants, dwarfs, armless men, the obese, the thin—you name them—they were there. Though a few of them were exploited, most of them were performers and entertainers who were applauded for having turned a potential liability into a profitable and valuable identity peg (Bogdan, 1988, p. 268). These human exhibits had no objection to being put on display; in fact, most of them enjoyed the attention. They were comfortable with what they were, and they did not believe that the word *freak* should be applied to them. By the beginning of the 20th century, the display of human oddities for amusement had fallen into disrepute in the United States. Part of the reason was the growing fear that these individuals might reproduce and transmit their physical traits to future genera- tions (Bogdan, 1988, p. 62). The other major factor was that after physicians organized into the American Medical Association in 1847, they worked to establish their claim to expertise in regard to more and more human conditions. Human exhibits such as those displayed in the freak shows came to be medical- ized as sick and in need of cure rather than as strange and curious (Bogdan, 1988, p. 66).

Some attributes of individuals are assessed in light of other attributes that they have (or are assumed to have); if they had a different cluster of attributes (or could find new relationships), then some of their discrediting (or discreditable) attributes would be viewed differently. Consider the following letter to "Dear Abby" from a "Happy Wife."

Dear Abby:

When our daughter was a baby, I found her pacifier in our bed. I thought it had dropped out of her mouth while she was in our bed, but later I found the pacifier in the drawer of our nightstand table, and I couldn't for the life of me figure out how it got there.

Then one morning I woke up early and saw my husband sound asleep with the pacifier in his mouth! We had a good laugh over it, and that evening when I fixed the baby's bottle I jokingly asked him if he wanted a bottle too. He said yes, so I fixed him one.

He loved it, so I kept fixing him a bottle right along with the baby's. I took the baby off the bottle when she was fourteen months old, but my husband still has one every night, and he is thirty-seven. Please don't use our names as my husband is well known here. He works on the space program. Thank you. (Van Buren, 1981, p. 175).

How can the husband *not* be a deviant? The wife is embarrassed to have her name printed, and both she and her husband seem to realize that most other fathers are not nursing on their babies' pacifiers and drinking from their bottles. Yet how can he *be* deviant? Both the wife and husband laugh over the husband's acts, and the wife really seems to experience very little consternation over her spouse's fondness for infant paraphernalia. Maybe other characteristics in the cluster of what he is and does (employee on the space program; good provider; good sense of humor) are sufficient for his wife to be tolerant of his less commendable traits. Audiences differ in what they accept or reject in others (Goode, 1997, p. 29).

One possibility for some people with a spoiled identity is to construct voluntarily a new identity or to move to a status that is not so discrediting. This option is not always available—not everyone can afford the cosmetic surgery to eradicate wrinkles on the face or to implant a full head of hair. However, some identities can be changed, and these transformations are in reach of practically anyone who wants to expend the time and effort to achieve them. However, movement to a nonstigmatized status does not necessarily resolve all the old identity problems. An individual may find that the stigmatized identity had offered a measure of comfort and that movement to a new status is not as rewarding as was hoped. Anticipated benefits fail to materialize, and the problems that were blamed on the former identity still exist. This situation can produce its own identity crisis (English, 1993, pp. 235-237).

Discrediting Lips

In November 1996, a 13-year-old girl named Karla Chapman did what she had done many times before—she went to class at the Runyon Elementary School in Pike County, Kentucky. This day, however, she was declared to be a distraction by the principal, Rosa Wolfe. Three times the principal warned her, and three times Karla Chapman defied the principal's authority. The problem? Karla Chapman wore *black* lipstick. At another time or place, with different people, this probably would have been no big deal. However, at this time and place, with these people, it developed into a very big deal indeed. Karla found that her choice of lipstick color (which she insisted was her business, not the school's), coupled with her refusal to wipe the stuff off as ordered, got her suspended for 3 days. On the day that her suspension ended, Karla Chapman got ready for school, donned a different colored lipstick, and arrived at the elementary school ready for classes to start. However, she was once again prohibited from entering because her new choice of lipstick color—dark purple—was still unacceptable to school personnel. The principal would brook no opposition, so Karla was put in a difficult situation. She had to return to school, or she would be considered a truant—which at Karla's age qualified her as a delinquent—but she could not return if she continued to wear lipstick that was considered distracting by the principal (Mueller, 1996).

It is difficult for an outside observer to this incident to understand what exactly about black lipstick makes it so distracting and what exactly made Karla Chapman such a problem to the principal. Certainly, nothing unique to black lipstick makes it any more upsetting than any other color lipstick. Even red lipstick could be distracting on a 13-year-old (depending, of course, on how it was applied). Yet a principal does have an obligation to keep the school setting calm and orderly. One wonders if the problem was not so much the black lips as the fact that a teenager would refuse to comply with orders to wipe the stuff off (and thereby challenge the authority of those who have the power to decide what is a proper or an improper appearance). If Karla Chapman had been born with black lips, the principal would have found herself in a more difficult situation in trying to expel her. Though Karla still would have been sporting a "distracting" color, the attribute would be widely viewed as something over which she had little or no control. The power and legitimacy of rule makers would not have been challenged directly by a young person's deliberate act of insubordination, and they could have afforded to be more gracious and understanding than they were in this particular case. This incident certainly shows that stigma is one possible outcome of the negotiations between people about the propriety of particular attributes.

Discreditable Tattoos

A tattoo has its own social functions and may be either a stigma symbol or a prestige symbol depending on a host of factors: who the tattooee is, what the tattoo represents, where the tattoo is placed, and the nature of the relationship between the tattooee and his or her significant others. Tattoos have been used to identify social outcasts and to make it more difficult for them to blend in with others. In Japan, in the sixth century, criminals were tattooed on the arms and face, and in the 1800s, convicts in correctional facilities in Massachusetts had "Mass S.P." and the date of their release tattooed on their left arms (Sanders, 1996, p. 376). Even in these cases, however, the tattoos were not automatically discrediting: It depended strongly on the audience that witnessed them. As with lipstick and other physical attributes, observers may have conflicting meanings over tattoos and how they should be evaluated.

A tattoo almost always means something important to the individual who gets one, and the decision to get a tattoo is almost always strongly influenced by one's view of self and the nature of one's social relationships (Mifflin, 1998, p. 27). One reason people get a tattoo is to change or to beautify their bodies. Beyond this rather general objective are more specific ones of getting a tattoo to disaffiliate from some groups, to affiliate with other groups, and to show one's independence and individuality. Getting the first tattoo automatically moves a tattooee into a new social category—people who have tattoos—and out of an old social category—people without them. A tattoo design—a club insignia, the name of a loved one, a symbol of one's hobbies—is selected for its information-conveying properties (although only the tattooee may understand what it represents) and its potential to bind the tattooee with others who have similar interests, experiences, or tattoos.

The stigma of a tattoo may isolate an individual from the conventional world. Most people with tattoos have been the focus of attention because of their markings. The actual or potential ridicule from others may be sufficient for a tatooed individual to seek contacts with supportive others and to be careful about how and to whom the tattoo is shown. This circumspection is especially likely if the other is an authority figure or has influence over the tattooee (Sanders, 1996, p. 371). People with tattoos classify and evaluate others on the basis of how these others classify and evaluate them. This "evaluation of the evaluators" helps people with tattoos to determine how accepting other people might be. Tattoos are used by their owners as a litmus test of social harmony between others and themselves.

Men will usually have a tattoo placed on an arm, and women will usually have one placed on a shoulder, the back, a breast, a hip, or the lower abdomen. Part of the variation in location is explained by the fact that tattooists charge less for a tattoo placed on the arm or leg, and part of the variation is explained by the fact that tattoos on the arms and legs are a bit less painful to get than tattoos on areas with a higher concentration of nerve endings. The stigma-producing potential of a tattoo also plays a role in where it is placed. For men, tattoos are less discrediting and are most often used to symbolize masculinity and toughness. For this reason, figures such as snakes, skulls, eagles, and bloody daggers are popular, and they are most often put in a conspicuous place. For women, however, tattoos are usually placed somewhere that can be easily covered because women realize the stigmatizing potential that tattoos have for them (Sanders, 1996, p. 369). It is both interesting and instructive that some people will purposely use a "tarnished" cultural product such as a tattoo to increase their feelings of self-worth and to transform their identities in the eyes of others (Sanders, 1996, p. 375).

Conclusions

Social dynamics are always involved in the construction of personal identity, and social context is always critical in understanding how personal attributes are viewed and the degree of stigma attached to them. Putting on lipstick or getting a tattoo reflects group variables, personal motivation, and intent, but its deviancy is in the eyes of the beholders and therefore variable and contextual. Any attribute can become an identity sign that people—self and others—use to categorize and to evaluate its possessor. Some of these attributes become the building blocks of master statuses, and some of these master statuses become the building blocks of spoiled identities. Individuals usually find that handling an identity that is discrediting—actually or potentially—requires a great deal of interpersonal work if they are to maintain their self-respect in the face of scornful reactions from others (Hughes & Degher, 1993, pp. 300-312).

Square Pegs and Round Holes:
Eccentrics and Eccentricities

Danielle Willis is a vampire, or so she claims. She sleeps by day, works by night (she's a fiction writer), and drinks human blood, partly for nourishment and partly because it excites her. She paid her dentist to install a permanent set of porcelain fangs over her incisors. She does not change into a bat and then bite hapless victims on the neck, however. What she does is use a syringe to extract blood from a willing partner (who she is confident has no blood-borne diseases) and then drink it, either right on the spot or at some later time. For Willis, the consumption of the bodily fluids of another is an expression of intimacy and trust. Hundreds of vampires such as Willis live throughout the United States, and some of them believe (or at least hope) that drinking blood ensures their immortality ("Interview With a Vampirette," 1997). Are these people real bats, just batty, or something else?

Eccentrics are quirky or even queer people who have thrown off the bonds of conformity and who pursue whatever wild hair intrigues them. Some eccentrics are successful people, and their eccentricities are just part of what they are. Other eccentrics have gained fame and fortune because of their eccentricities, and their eccentricities shaped their lives and their identities (Nash, 1982, p. xix). Still other eccentrics are abysmal failures at practically everything they do, partly because they are obsessed with their eccentricities (Wallace, 1957, p. 11). If an individual has the forbearance, ability, or good luck to triumph in some field, his or her eccentricity is likely to be overlooked or even admired. However, if an eccentric fails to gain prominence in some valued field of human endeavor, then his or her oddness is more likely to be disturbing to others, and he or she is more likely to be condemned or ridiculed (Wallace, 1957, pp. 11-12).

The label of *eccentric*—like all labels—is relative, and the ground rules for what makes one eccentric change all the time. Alexander Wortley had a deep suspicion of zippers in men's trousers, so he removed them from any pair he purchased. The reason? He did not want a lightning conductor so close to such a sensitive body part (Adams & Riley, 1988, p. 217). Wortley's anxiety over genital shocks is not at all unreasonable; it is how he went about protecting himself from electrocution that seems strange. The Reverend George H. Munday was a renowned Philadelphian preacher of the 19th century. Parishioners gathered by the hundreds to hear his sermons but mostly to observe his odd trait: He refused to wear a hat at a time when all male Quakers did (Sifakis, 1984, pp. xvi-xvii). Joseph Palmer (1788-1875) moved to the city of Fitchburg, Massachusetts, in 1830. Some of his new neighbors shunned him, and others threw rocks at him (and at his house). Businesspeople refused to cater to him, and religious people prayed for his redemption. Women avoided and feared him, often crossing to a different side of the road when they saw him coming. What was Mr. Palmer's problem? He was one of the first individuals in the United States to grow a long beard. He was so upsetting to others that one day four men armed with scissors and a razor attacked him, threw him to the ground, and tried to shave him forcibly. Through all his trials and ordeals, he steadfastly kept his beard. His gravestone in Evergreen Cemetery in Leominster, Massachusetts, tells the story. It reads, "Persecuted for wearing the beard" (Sifakis, 1984, pp. 69-70).

Some eccentrics are what they are because accidents of birth made their pursuit of novelty more likely. Michel Lotito, a Frenchman from Grenoble, has an amazing ability. He can eat practically anything. When he was 16, he was drinking mint tea with friends at a French cafe, and the rim of the glass accidentally broke off in his mouth. Instead of spitting the piece out and complaining about the defective glass, he chewed the piece up and swallowed it. Because he experienced no adverse effects, he soon realized that he had a special talent (Flaherty, 1992, pp. 129-130). He went on to become a professional entertainer whose performance consists of eating his way through things such as television sets, aluminum skis, supermarket carts, bicycles (he likes the chain the best), razor blades, coins, glasses, bottles, bullets, and phonograph records. He has even eaten an entire airplane—a Cessna two-seater—piece by piece. Lotito cuts objects into bite-size pieces, lubricates his digestive tract liberally with mineral oil, and drinks lots of water as he eats the debris. Surprisingly enough, though he can eat stuff that would kill an ordinary person, he has difficulty digesting bananas and eggs (Adams & Riley, 1988, p. 231). Lotito was awarded a brass plaque by the *Guinness Book of World Records* to commemorate his eating eccentricities. He was so honored that he ate it (Flaherty, 1992, p. 130).

David Weeks and Jamie James (1995) were interested in studying eccentrics, but where does a researcher go to find them? They decided to place a terse statement on index cards—"Eccentric? If you feel that you might be, contact Dr. David Weeks at the Royal Edinburgh Hospital"—and then to post the cards in places such as libraries, supermarkets, and universities throughout the Edinburgh area (Weeks's phone number was included on the cards). British journalists eventually learned of the study, and their stories helped to publicize the research. The study was also discussed on television and radio. The U.S. press finally got wind of the research, and that allowed Weeks and James to study eccentrics in more than one society. When the smoke had cleared and the difficult process of pruning the sample to find "true eccentrics" was completed, the authors had 789 potential eccentrics in their study, 309 men and 480 women. The authors interviewed their sample of eccentrics, gave them personality evaluations and IQ tests, and got psychiatrists to examine them for evidence of mental illnesses. All the information was used to construct a "group portrait" of eccentrics.

One of the most irrepressible traits of eccentrics is that they pay little attention to the derision and ridicule that they often receive from others (Weeks & James, 1995, p. 16). In some cases, they are unaware that they are upsetting or annoying to others, but most times they are just unconcerned with what others think. Weeks and James found some common traits in their eccentrics: (a) nonconforming; (b) creative; (c) curious; (d) idealistic; (e) happily involved with one or more quirky hobbies or obsessions—called hobbyhorses by the authors—usually five or six; (f) felt different from others (and they had since childhood); (g) intelligent; (h) dogmatic; (i) independent; (j) unusual in living and eating practices; (k) isolated from contacts with others and indifferent to their opinions; (l) having a mischievous sense of humor; (m) single; (n) being either the oldest child in a family or an only child; (o) having poor spelling ability. The first 5 were found in *every* eccentric in the study, and these 15 range from the most frequently found to the least. Some of these character traits are little more than synonyms for *eccentric*—nonconforming, creative, obsessive, opinionated, having peculiar living arrangements or eating habits—and others—isolated, unmarried, independent—are probably a response to how others have reacted to the eccentrics.

Eccentrics do occupy an indeterminate status. They are fascinating to others (and may even be a source of envy) while also upsetting to them. Eccentrics believe that they are right in what they do, and they are not usually unhappy with their unconventionality (Weeks & James, 1995, p. 16). They tend to do exactly as they please, and they are usually unconcerned with what is proper or with what others want them to do. "Eccentrics are people who take boundless joy in life, immoderate men and women who refuse to violate their ideals. Their minds are always buzzing furiously with ideas. . . . At the root of eccentricity is a healthy and determined irreverence" (Weeks & James, 1995, p. 254). The eccentric's unbridled freedom and independence of thought and action (or is it irresponsibility or some slavish obedience to the goal of being weird?) may rub others the wrong way. What gives eccentrics the right to do what they want when the rest of us cannot?

Eccentrics may actually be more alarming than other kinds of deviants. To be sure, the eccentricities of eccentrics rarely break the law, but this does not mean that eccentrics are simply laughable buffoons, providing the spice of life. Whereas the typical deviant probably does know the difference between right and wrong and does not flaunt or challenge the rules openly, the eccentric seems to be out of touch with the ordinary concerns of the ordinary folk. Not only is the eccentric odd in the eyes of the beholders, he or she goes to go to great lengths to be different and separate from others, defends his or her oddness as perfectly proper, is indifferent to the expectations and wishes of others, and appears unable or unwilling to understand why others would be upset by him or her. Whatever tensions might exist between individual desires and the conventional forces of social control have been resolved successfully by the eccentric in his or her own interests (Suran, 1978, p. 206).

For most of us, some feature of what we do or what we are—our personal identities—could be seized on and branded as eccentric by others. These eccentricities could then become part of a master status and the principal way that others view and respond to us. A process of retrospective interpretation, coupled with selective perception, could seal our fate, and each one of us could become the core deviant of some particular group at some particular point in time.

Summary and Conclusions

This chapter has shown that attributes themselves are a source of classification and judgment. People are stigmatized and bedeviled for things over which they have little control. "Normal" and "stigmatized" are perspectives on persons, so we always will need a language of relationships to discuss any discrediting or discreditable attribute such as physical appearance, body modifications such as tattoos, or eccentricities. Deviance can easily evolve into a master status and be accompanied by shame and embarrassment. Retrospective interpretation is a dynamic process in which a person's social identity is reconstituted in the eyes of others.

References

Adams, S., & Riley, L. (1988). *Facts and fallacies.* New York: Reader's Digest Association.

Becker, H. (1963). *Outsiders: Studies in the sociology of deviance.* New York: Free Press.

Bogdan, R. (1988). *Freak show: Presenting human oddities for amusement and profit.* Chicago: University of Chicago Press.

Bowditch, C. (1993). Getting rid of troublemakers: High school disciplinary procedures and the production of dropouts. *Social Problems, 40,* 493-509.

Buss, D. (1994). *The evolution of desire: Strategies of human mating.* New York: Basic Books.

Curra, J. (1994). *Understanding social deviance: From the near side to the outer limits.* New York: HarperCollins.

English, C. (1993). Gaining and losing weight: Identity transformations. *Deviant Behavior, 14,* 227-241.

Flaherty, T. (1992). *Odd and eccentric people.* Alexandria, VA: Time-Life.

Garfinkel, H. (1956). Conditions of successful degradation ceremonies. *American Journal of Sociology, 61,* 420-424.

Goffman, E. (1963). *Stigma: Notes on the management of spoiled identity.* Englewood Cliffs, NJ: Prentice Hall.

Goode, E. (1997). *Deviant behavior* (5th ed.). Upper Saddle River, NJ: Prentice Hall.

Gregersen, E. (1983). *Sexual practices: The story of human sexuality.* New York: Franklin Watts.

Hughes, E. C. (1945). Dilemmas and contradictions of status. *American Journal of Sociology, 50,* 353-359.

Hughes, G., & Degher, D. (1993). Coping with a deviant identity. *Deviant Behavior, 14,* 297-315.

Interview with a vampirette. (1997, February 9). *New York Times Magazine,* p. 17.

Kitsuse, J. (1962). Societal reactions to deviant behavior: Problems of theory and method. *Social Problems, 9,* 247-257.

Kitsuse, J. (1980). Coming out all over: Deviants and the politics of social problems. *Social Problems, 28,* 1-13.

Mifflin, M. (1998). Written on the body: Women and the art of tattooing. In L. Salinger (Ed.), *Deviant behavior 98/99* (3rd ed., pp. 26-28). Guilford, CT: Dushkin.

Mueller, L. (1996, November 19). Father of black-lipsticked girl arrested. *Lexington Herald-Leader* (Lexington, KY), pp. A1, A8.

Nash, J. R. (1982). *Zanies: The world's greatest eccentrics.* Piscataway, NJ: New Century.

Ridley, M. (1993). *The Red Queen: Sex and the evolution of human nature.* New York: Penguin.

Sagarin, E. (1975). *Deviants and deviance: An introduction to the study of disvalued people and behavior.* New York: Praeger.

Sanders, C. (1996). Getting a tattoo. In E. Runbington & M. Weinberg (Eds.), *Deviance: The interactionist perspective* (6th ed., pp. 364-377). Boston: Allyn & Bacon.

Schur, E. (1971). *Labeling deviant behavior: Its sociological implications.* New York: Harper & Row.

Sifakis, C. (1984). *American eccentrics.* New York: Facts on File.

Small, M. (1995). *What's love got to do with it? The evolution of human mating.* New York: Anchor.

Suran, B. (1978). *Oddballs: The social maverick and the dynamics of individuality.* Chicago: Nelson Hall.

Trost, C. (1982). The blue people of Troublesome Creek. *Science, 82*(3), 34-39.

Van Buren, A. (1981). *The best of Dear Abby.* Kansas City: Andrews & McMeel.

Wallace, I. (1957). *The square pegs: Some Americans who dared to be different.* New York: Knopf.

Weeks, D., & James, J. (1995). *Eccentrics: A study of sanity and strangeness.* New York: Villard.

3

Sexual Diversity

Introduction: Pigeonholing People

So little is known about what people around the globe and in past societies actually do or have done in bed (or in hut, or in tent, or in igloo, or in cave, or in the big outdoors) that *any* statements about human sexual variance are speculative. Books that purport to tell us what really happens in our sexual lives really do not. They offer little more than biased information, coupled with a few anecdotes, about those sexual experiences that some people were willing to share with researchers (Carter & Sokol, 1989). These "pictures" or "portraits" of human sexuality are ripped from the social context that defines them, so they are stripped of much of what makes them sexual in the first place. Even if the respondents are being honest, the data are still not representative enough to give anything close to a satisfactory picture of the nature, diversity, and relativity of human sexual experiences. If anything natural or normal can be found in human sexuality, it is that we as a group find a great many things sexually arousing and that our sexual identities are subject to a great deal of flexibility and variability.

People who are developing their sexual identities confront a great deal of ambiguity and confusion. It is quite likely that shared sexual identity is not identical to shared sexual experience, and people pigeonholed in the same category may have vastly different experiences, outlooks, and temperaments. Similarities and dissimilarities—homogeneity and heterogeneity—do not lie so much in actual experiences as they do in how these personal experiences are conceptualized and interpreted (Rust, 1992, p. 381). Gender is not a natural, biological fact (Moon, 1995, p. 496). It is an ongoing and dynamic social construct that is influenced by environmental learning, individual interests and capabilities, and intricate presentations of self through both physical and symbolic cues (Tewksbury, 1994).

The Curious Case of the Berdache

Some Native American groups offered their members a sex status in addition to male or female, usually called *berdache*. Usually, a berdache was an anatomical male who combined parts of the masculine and feminine gender into a unique role (Williams, 1986, p. 142). Berdaches moved freely between men's and women's groups because of their unique status, and berdachism was not always permanent. Some men were berdaches for awhile but eventually changed and acted like regular men (Williams, 1986, p. 78). Transgender statuses such as that of the berdache show us quite clearly that the facile division of the human race into two mutually exclusive categories of male/female or masculine/feminine fails to convey adequately the full range of possibilities with regard to sex, gender, and sexuality.

Attitudes toward the berdache went beyond mere recognition and acceptance to attitudes of reverence or even awe:

> American Indian cultures have taken what Western culture calls negative, and made it a positive; they have successfully utilized the different skills and insights of a class of people that Western culture has stigmatized and whose spiritual powers have been wasted. (Williams, 1986, p. 3)

Berdaches were respected because they had that double vision that often flows from occupying a go-between position between male and female, between the physical and the mental, and between the spirit and the flesh (Williams, 1986, p. 30). They were able to transcend the confines of a one-gender worldview, and they were respected for it. Their secure position in the social life of the community practically ensured that they would enjoy both personal and social success (Williams, 1986, p. 57).

Early Native American groups were some of the most egalitarian on earth. This sexual equality meant that females were of great importance to the life of the community and that their contributions—social, economic, political—were every bit as valuable as those of males. "Because women had high status, there was no shame in a male taking on feminine characteristics. He was not giving up male privilege or 'debasing' himself to become like a woman, simply because the position of women was not inferior" (Williams, 1986, p. 66). Where women have high status, no loss of esteem accrues to a male who moves in a feminine direction. If anything, it may indicate special qualities and be a source of pride and high esteem.

It would be incorrect to conclude that the berdache status was universally respected and that berdaches were uniformly honored. Would we really expect this given the diversity of human relationships? Some of the enmity directed at some berdaches by their fellows reflected the impact of Europeans and their Christian ideals on Native Americans. These early Europeans tended to be intolerant and inflexible when confronted with any sexual practices or gender roles that they defined as unusual or different. The persecution of homosexuals and sodomites made a great deal of sense to these visitors to the New World. It gave them one way to establish, in their own minds at any rate, their superiority over other people, while convincing themselves that their way was the only natural and proper way. By the 13th century in Europe, the suppression of homosexuality was well underway, and by the 14th century, male sodomy was viewed throughout Europe as a capital crime. Sodomites, homosexuals, and pederasts were lumped together with heretics, Jews, and lepers, all of whom were branded as extreme threats to all good Christians. The "abominable sin" of sodomy was suppressed wherever it was found (Moore, 1987, pp. 93-94).

When worlds collide—or when representatives of different cultural belief systems come into contact with each other—the conflict that this entails is fertile ground for the creation of deviance. It is an old story. Members of some in-group find something that an out-group does that the in-group does not, and they then use it as a basis for invidious comparison. This process gives full rein to ethnocentric feelings, and it becomes a way for in-group members to devalue, to persecute, and, if possible, to eradicate the out-group. Once the condemnation of outsiders gets started, it has a snowballing effect, and it becomes easier to attack *anyone* who is different.

When the Spaniards came to the Americas, they carried their prejudices with them. Rather than accepting Native Americans that they met for what they were, they used their interpretation of the Christian Bible to justify the extermination of the indigenous population. It was not too hard for them to decide—they had much to gain—that the pillage and plunder, arson, murder, rape, and all the rest were an acting out of God's will and their road to personal salvation (Williams, 1986, p. 137).

Native American cultures are not the only ones to have a legitimate, valued gender role in addition to masculine and feminine. Societies in many parts of the globe—Oceania, Africa, India, Siberia, Asia, Australia—have a gender category comparable to the berdache (Williams, 1986, pp. 252-275). What distinguishes these places from the West is not that we in the West are more normal but that we in the West provide no legitimate, institutionalized, valued status for gender-benders such as the berdache to occupy. In Western cultures it is very difficult for an individual *not* to embrace one of the two proffered, mutually exclusive categories of male/female, masculine/feminine, or gay/straight even if he or she feels uncomfortable doing so (Weinberg & Williams, 1994).

The Evolutionary Setting of Sex, Gender, and Sexuality

One view of the relationship between evolution, sexual identity, and sexuality gives genes *the* preeminent position. According to this approach—going by names such as *evolutionary biology, evolutionary psychology,* or *sociobiology*— practically all sexuality, human and nonhuman, is designed to increase the reproductive success of an individual creature, and anything that makes repro-ductive success more likely will succeed at the expense of anything that does not (Ridley, 1993, p. 20). This approach generates a rather dismal appraisal of human sexual experience. Men, the story goes, try to acquire power and resources and then use them to attract women who will bear them offspring to ensure that their male genes are passed on to future generations. Simply, wealth and power are the means to attaining women, and women are the road to genetic immortality. The story continues that women search for good husbands with abundant re-sources who will be able to offer support and protection for them and their children. Women are portrayed as genetically determined to be more selective and choosy in selecting mates and more faithful to them, and men as preferring to spread their genes around (Buss, 1994, pp. 19-48).

Human sexual customs are no doubt related to human biological charac-teristics, but exactly how is a matter of dispute. It is customary to assert that biological factors are the cause and that external social or sexual relationships are the effect or outcome, but this direction of causality may be all wrong. Sexual relationships and human experiences do change the human body in all kinds of ways. When men are actively involved in sexual relationships, their testosterone levels increase dramatically, much more than if they achieve sexual pleasure through solitary acts of masturbation (Kemper, 1990). Men may actually have altered sperm counts in response to the possibility that their female mates are being unfaithful (Small, 1995, p. 118). Ridley (1993) declared that it is the choosiness of humans in selecting their permanent mates that has forced the human brain to expand because wit, virtuosity, inventiveness, and individuality are all sexually appealing (p. 344).

Theories that suggest that gender and sexual identity are caused by brain differences are almost always based on the idea that the left and right hemispheres develop differently in boys and girls. It is usually hypothesized that the male brain is more lateralized (the hemispheres are specialized in their abilities), whereas the female brain is more symmetrical because her corpus callosum (the part of the brain that joins the two sides of the brain) is larger and contains more fibers. So, the story goes, women have better interconnected brain hemispheres than men, which is used to explain why women excel in talk, feelings, intuition, and quick judgments.

It may be true that males and females do have different brains and that these differences account for male/female differences. However, the research on brain differences between males and females is often sloppy, based on exceedingly small samples, and constructed to reaffirm preexisting stereotypes about sexual differences. The fact is that our sexual behaviors are far more complex than the simple distinction between right brain and left brain would imply (Tavris, 1992). Men and women are much more alike than they are different, and the range of differences *within* a sexual category (male or female) is always greater than the range of differences *between* a given male and a given female.

This interest in possible brain differences between males and females has spilled over into the study of brain differences between heterosexuals and homosexuals. Is there a homosexual brain? Simon LeVay (1993) conducted some suggestive research. He autopsied the hypothalamuses from 19 homosexual men who had died from AIDS and 16 heterosexual men, 6 of whom had died from AIDS. Six women were also autopsied, but their sexual orientation was unknown. He discovered that his male specimens had a thickening in brain cells not found in females and that heterosexual men had thicker cells than homosexual men. LeVay concluded that sexual orientation is directly related to brain structure. However, that conclusion may be unwarranted. The differences in brain structure that he found may simply have been caused by the impact of the AIDS virus on brain chemistry and not the impact of brain structure on sexual orientation (Byne, 1994, p. 53).

Nobody has yet provided an incontrovertible explanation for how something as complex and dynamic as sexual orientation could be biologically determined.

Though certain biological markers must be present for an individual to be sexual, sexual orientation and gender have strong social components. The most reasonable view at this time is that sexual orientation has prenatal and postnatal causes that work together to produce an orientation that gets more stable over time until it reaches a point of being practically unchangeable (Money, 1988, p. 4). Heterosexuality and homosexuality are not absolutes. They are different ends of a continuum, and people can fall at different places on that line depending on a host of factors (Weitz & Bryant, 1997, p. 40).

Sexual Differentiation and the Double Standard

Sexologists in the United States in the early 20th century crafted a unique view of sexual differentiation. They took it for granted that the world was automatically and naturally divided into two mutually exclusive categories: masculine and feminine, or male and female (Birken, 1988). This view that a two-category sexual differentiation was both natural and normal offered them abundant riches. It allowed them to seize control of the sex-gender-sexuality complex and to decide what was normal and what was not. The opportunity to define/invent problems and then to offer "fixes" for them is critical to any newly emerging profession. The biomedical approach to sexual variation, with its appeal to innateness and naturalness, offered advantages to early sexologists that a more relativistic approach never could (Irvine, 1990, p. 285). We must never forget that *any* decisions about what is normal sexuality and what is a normal sexual being always reflect historically determined, culture-bound understandings about the proper ends of human social and sexual relationships. Ridley (1993) insisted that men and women have different human natures. Women's minds, he claimed, evolved for the purposes of bearing and rearing children and to gather food; men's minds evolved for the purposes of achieving upward mobility in a male hierarchy, fighting over women, and bringing home the bacon (i.e., meat) to a starving family (Ridley, 1993, p. 248). He insisted that these differences between the sexes are a product of evolution, so that no amount of wishing or propagandizing will make them less real (p. 276).

Ridley was aware of the potential for harm that his ideas presented, but he had a strategy for protecting potential victims. Difference, he insisted, is not inequality. Preferences and characteristics of one sex are not inherently better than those of the other sex, and the differences between the sexes should be neither denied nor exaggerated (p. 276). His claim that males and females are naturally different but still naturally equal is unsatisfactory. The universal tendency to pigeonhole is almost always coupled with the universal tendency to prize some characteristics and to condemn others. Social differentiations are almost always transformed into social evaluations, and social evaluations translate into social deviance. It would be nice if this did not happen, but it almost always does. More often than not, sexual asymmetry becomes part of an ideology in which one group maintains its dominance by directly subordinating or subjegating members of some other group (Kimmel, 1996; Spain, 1992).

Some women in the world today on their wedding nights must (*absolutely* must) possess that membrane known as a hymen (Saadawi, 1982). If this thin membrane that covers the opening of the vagina was missing, it would automatically call into question a woman's status as a virgin. Therefore, it is essential that the bride's hymen be capable of being ruptured and then bleeding profusely onto a bed sheet or white towel. The bloody cloth is then paraded around by the father of the bride or the proud husband to show the throng of relatives, friends, and well-wishers that all is well. The bride has been chaste and pure, and the family honor is preserved.

A female who lacks a hymen or who has one too strong for it to be easily broken—many reasons other than premarital sexuality exist for the lack of a hymen in a female—will have to resort to trickery: planning her marriage so it will coincide with her menstrual period or secreting a small container of chicken blood near the opening of the vagina to guarantee blood flow at the proper time. In countries with a hymen-breaking custom, a male's status is less dependent on his own moral behavior than it is on that of female members of his household. A man can act in the most scurrilous of ways and not lose face, but sexual experimentation by a woman is viewed as disgraceful to herself and to her family. More important, it lessens her value as a prospective bride.

The custom of hymen breaking pales by comparison with another custom that reflects the same double standard. In some places on earth, every female child undergoes an extreme and traumatizing mutilation of her genitals as a matter of custom. It is known as circumcision, genital alteration, or—the most accurate term—female genital mutilation. In some regions of the world, the clitoris is amputated, an operation called a clitoridectomy (Saadawi, 1982, pp. 7-11). In other regions—the Sudan in North Central Africa, for example—the procedure is more radical. All the genital organs—clitoris and portions of the labia majora and the labia minora—are cut away with a knife, a razor, a pair of scissors, or even a sharp rock. The remaining flesh is then sewn together with a sharp needle, and the female's legs are bound together for many weeks to allow the skin to heal. A small hole for the passage of urine and menstrual blood is all that remains of the vaginal opening (Burstyn, 1997, p. 21). On her wedding night, sometimes years after the mutilation, her husband cuts her vagina open with a knife to accommodate sexual intercourse. If a pregnancy results, she will be cut open even wider to allow her child to be born. After the delivery, the woman's vaginal opening may be sewn together once again to ensure her fidelity to her husband and to make her vagina tight enough that her husband will enjoy himself during sexual intercourse (Ziv, 1997, p. 11).

The custom of female genital mutilation is kept alive partly by the belief that terrible consequences will befall a female who forgoes the procedure. Some people believe that an uncircumcised clitoris will grow so large that it will drag on the ground or that uncircumcised females will be unruly, oversexed, unclean, childless, and unmarriageable (Burstyn, 1997, p. 20). The Bambara of Mali believe that a man will die if his penis comes in contact with the clitoris during sexual intercourse, and in Nigeria some groups believe that a child will die if his or her head brushes against the mother's clitoris during delivery (Mackie, 1996, p. 1009). All of these beliefs are, of course, unfounded, but the people who practice genital mutilation are caught in a belief trap: No one is brave enough to risk the costs of failing to perform the operation (Mackie, 1996, p. 1009).

The real costs of having the features of one's anatomy defined as unacceptable fall most heavily on the female who is mutilated. It is usually done between infancy and adolescence, most often on a girl between ages 2 and 9. The young girl is brutally grabbed and held down by several adults until the cutting is complete. She is then sewn together and bound for several weeks. Beyond the betrayal this evidences to the child is the possibility of infection, infertility, and even death. The mutilation practically guarantees that she will never be able to experience sexual pleasure. Africa is the region where the custom is most frequent, but it is also found in parts of South America, Southeast Asia, and the Middle East (Ziv, 1997, p. 10).

Sexuality—A Sociocultural Understanding

Human sexuality does not take place in a vacuum. It is a very social experience, and this feature is why it is so wrong to exaggerate the role of biology and to focus on the individual alone. "Sex involves negotiation and interplay, the expectation and experience of compromise. There is competition; there is coop-eration" (Laumann, Gagnon, Michael, & Michaels, 1994, p. 5). Sexual behavior is learned in the context of gender identity, and it represents an outgrowth of cultural and psychological factors (Gagnon & Simon, 1970, p. 12). We need a sociocultural perspective on human sexuality to provide some balance to explanations of human sexuality that are too biological and deterministic (Reiss, 1986, p. 1).

No important and enduring human social relationship is going to be ignored by human beings and left entirely to chance or luck. Humans are going to judge it, deliberate on it, worry about it, like it, hate it, praise it, condemn it, modify it, chase it, avoid it, or justify it, as the case may be. At some point, human relationships will be scripted by certain groups, and these scripts will be incorporated into their cultures. This is certainly true of human sexuality. Sexual scripts are constructed to tell individuals the appropriate and inappropriate ways of acting, thinking, and feeling with regard to things sexual. Each social group will usually have a main sexual script known to most members (Gagnon & Simon, 1973, p. 19), but some sexual scripts are familiar to a relative few (Laumann et al., 1994, p. 6). With sexual deviance, as with perhaps no other form of social activity, we see that people go beyond the prevailing cultural scripts and invent their own (Tewksbury, 1996, p. 4). Practically any act that brings pleasure, no matter what it is, and no matter whether it is done all alone or with others, can be influenced by sexual scripts.

As valuable as the idea of sexual script is in helping us understand the social construction of sexuality, it is clearly not enough. We must remember that people construct actions on the basis of the consequences of those actions, using their definitions of situations (which are strongly influenced by their relationships to others). Human sexuality always involves interpretation, negotiation, anticipation, opportunity, and accessibility. The existence of sexual freedom means that a great deal of confusion exists over just what sexual behaviors should be accepted and tolerated and to what degree (Kelly, 1994, pp. 376-377).

Our sexual relationships reflect other features of society and culture, features that at first blush seem to have little to do with sexual relationships (Schur, 1988, p. 199). Some of these sociocultural elements do make sexual experiences personally rewarding and self-enhancing. Other sociocultural elements, however, may spoil our sexual relationships and make them far less meaningful and rewarding than they could be (Schur, 1988, p. 76). Dominant themes in American society—practicality, competition, individualism, consumerism, aggression, superficiality—interpenetrate the American way of sex and have promoted an impersonality and emotional emptiness in intimate relationships (Schur, 1988, p. 10). These factors conspire to increase the chances that sexuality will be too impersonal, coercive, and self-centered (Schur, 1988, p. xii). Kept within limits, the changes in intimate relationships may not be entirely bad. Because intimate relationships can now be separated from reproduction and other conventional restraints, it is possible for people to use intimate relationships exclusively to satisfy their emotional needs for closeness and self-disclosure (Giddens, 1992).

The new culture of intimacy legitimated sex as an arena for the achievement of pleasure, self-disclosure, and meaningful interaction (Seidman, 1991, p. 155). These objectives became important in their own right, separate from the achievement of other worthy goals such as reproduction, marriage, or love.

> Critics of the American intimate culture have not sufficiently appreciated that while expanded sexual freedom and a culture of eroticism may have some costs, the benefits—including expanded freedom and diversity, new sites of pleasure and self-expression and the creation of new identities and communities—are considerable. (Seidman, 1991, p. 194)

Sex evolved into a social cement that created new connections between people who might never have otherwise gotten together, and individuals were freer either to grow together or to leave unsatisfactory relationships (Giddens, 1992). Members of the majority branded some of these new unions as perverted, and their casualness (apparent or real) was used as the principal justification for condemning otherwise dissimilar pursuits. Casual sex was viewed as a problem, partly because it was casual and partly because it was associated with other troubling things such as illegitimacy, teenage pregnancy, irresponsibility, AIDS, STDs, marital instability, divorce, hedonism, and drug use.

The Myth of the Universal Turn-On

Because human sexual deviance has so much overlap with human sexuality, much of what is true about sexuality is also true about deviant sexuality as well. No intrinsic differences need exist between sexual deviance and sexual nondeviance. All that is required is that some group ban certain kinds of sexuality and bedevil those individuals who participate in the forbidden acts. Though most sexual deviance has no specific procreative function (sexual deviances may be a prologue to reproductive acts), this fact is not enough to account for its deviancy. Sex for pleasure is an integral part of human sexual experiences, and even nonhuman primates seem to participate in sexuality for purposes other than multiplication and division.

All people are not turned on by all things, and it is highly improbable that humans could learn to be turned on by everything. Undoubtedly, biology does influence sexual expression. No amount of cultural conditioning and socialization is going to make it possible for a human to orgasm indefinitely. Physical limits exist to how many sexual episodes one can have in a day, a year, or a lifetime. We can predict that some things are more likely than other things to turn people on, and we will be right more often than not. However, if they choose it, humans have access to an intricate world of sexual experiences, sexual objects, and sexual activities that is unique because it offers so much diversity and variation (Chodorow, 1994). Biological sexual urges are harnessed and amplified by social factors and put to a variety of social uses (Gagnon & Simon, 1970, p. 20). Among humans, practically anything goes because of the nature of their social relationships and cultural scripts, not because of the nature of their genes and biological equipment. Goode and Troiden (1974) made a valuable observation over 20 years ago that is still ignored by far too many:

> Sexual behavior is not dictated by the body, not by our animal chemistry, but by the mind, by our human relationships, by civilization, by what has historically come to be accepted as good or bad—in short, by convention. We do what we do in bed because we have learned to do so. And what we do *could* have been otherwise; by growing up surrounded by different customs, we would have been completely different sexual beings. At birth, the possibilities for what we could do in bed—or, for that matter, out of bed, or anywhere else—are almost boundless. (p. 14)

These observations—about convention, learning, human relationships—are as true today as they were when their book was first published.

The array of human sexual experiences is vast, especially when compared to that for nonhuman creatures, and the variation from culture to culture (as well as from subculture to subculture) is great enough that we must recognize a great deal of flexibility in human sexual experiences that surpasses that for *all* nonhuman animals, even more sexually playful ones. Human sexuality is *distinctly* human (Goode & Troiden, 1974, p. 13). A major reason is that human sexuality is constituted by feelings, thoughts, and the exchange of communication in ways that nonhuman sexuality never could be.

Some sexual turn-ons—such as klismaphilia (sexual arousal from enemas), urophilia (sexual arousal from urine), or coprophilia (sexual arousal from feces)—are quite restricted, being found among a relatively small number of individuals who may be suffering from severe disorders. However, other types of sexual deviance are practiced in a group setting and are associated with specific social networks, motives, rationalizations, and recruitment techniques to ensure the viability and continuity of the subculture of sexual deviance. The possibility for more group-based sexual deviance is even greater now because of a need for low-risk sexual outlets. Voyeurism (sexual arousal from viewing forbidden things), for example, has a variant in which a number of voyeurs will get together to spy on people. Sometimes they use telescopes, and they may even be viewing people who know that they are being watched (Forsyth, 1996, pp. 286-291).

Some individuals may become part of what Gregersen (1983) called the "chastity underground": people who have an interest neither in sexual activities nor in forming sexual relationships (p. 181). The scanty information available suggests that these people are not generally subject to any sexual dysfunction such as impotence or frigidity; they simply have a marked disinterest in sexual things. The authors of *Sex in America* (Michael, Gagnon, Laumann, & Kolata, 1994) reported that approximately one out of three women and one out of six men report a lack of interest in sex (for at least one of the 12 months before data collection); one woman in five and one man in 10 report that sex gives them no pleasure (p. 126).

Humans are continually finding new sexual turn-ons, using old body parts in new ways, finding new sexual paraphernalia to increase their sexual pleasures, establishing new relationships that provide new pleasures in the old ways, enjoying solitary sex, or finding comfort in no sex at all (temporarily or permanently). Humans may not have yet imagined the boundaries of their sexual potential, let alone exhausted all the possibilities. Alfred Kinsey, a biologist who eventually directed his attention to surveying and describing human sexuality, was more aware of the diversity of human sexuality than are many of his successors. He knew that practically any stimuli could be sexualized by someone (Kinsey, Pomeroy, Martin, & Gebhard, 1953). How else can we explain the fact that for some people riding horses is sexually arousing and for others smelling the exhaust fumes of automobiles is?

Bullough's (1976) comprehensive exploration of sexual variance throughout history makes it quite clear that wherever human sexuality is found, so is human sexual deviance and that what qualifies as sexual deviance is not uniform from place to place and time to time:

> Some [societies] emphasized chastity and tolerated intercourse only in a marriage relationship with procreation specifically in mind, others had a double standard in which almost everything was tolerated for the male but not for the female, and still others tolerated, if not encouraged, almost any sexual activity, there being all kinds of variation in between. (p. 19)

With sexuality, a vast underworld of hidden sexual practices exists that never becomes part of the official world. Even kissing, a valued and necessary part of lovemaking in some cultures, is defined as aberrant and unnatural in others (Tiefer, 1978, pp. 29-37). What one group defines as the epitome of sensuality, sophistication, and normalcy, another group brands as obscene, dangerous, and in need of immediate eradication (Beisel, 1993, p. 148).

Rigid rules of sexual conduct, coupled with punitive sanctions, probably have the primary effect of causing individuals to talk about their sexuality less and less and to conduct their sexual affairs with greater caution and secrecy. It is far less likely that a valued and enjoyable sexual practice will be abandoned entirely, no matter how harsh the penalties. The restrictions on sexual expression in 19th-century U.S. culture appeared inviolable. Chastity before marriage and fidelity after it were cultural imperatives. Sexual moderation was expected, and sexual intercourse was officially allowed only if it led to the birth of a child. Sexual practices such as masturbation, oral sex, and anal sex were strictly forbidden and rarely discussed. If any forbidden acts did occur, they had to be done very discreetly and carefully (Gay, 1986, pp. 201-202). Apparently, the Victorians did enjoy a wide range of sexual pleasures, and they privately rejected the ideal culture that called for a monotonous and unemotional sexual life (Lystra, 1989, pp. 101-102).

Kinsey's studies of sexuality (Kinsey, Pomeroy, & Martin, 1948; Kinsey et al., 1953) indicated that the sexual activities of Americans were far more flexible and diverse than ideal culture permitted and than most people suspected. By the middle of the 20th century, ideal culture itself started to change. More was learned about the sources of sexual pleasure, and conception could be separated from the sex act because of the availability of effective methods of contraception. The double standard started to erode as women left their traditional position as caretakers of home and children to seek their fortunes in jobs and work for advanced degrees in colleges and universities. Homosexuals and prostitutes became better organized and more insistent that they be left alone because their sexuality was their business and nobody else's (Bullough, 1976, p. 635). People felt freer than they ever had to embark on their own journey into sexuality and to use their imaginations to discover new sources of sexual pleasure with little concern for what prevailing sexual standards dictated. People became confident enough to divulge that they regularly violated prevailing sexual standards some of the time. The acclamation of sexuality as a pleasurable and important human relationship for both males and females went a long way toward legitimating the kinds of sexual experimentation—group and solitary—that would have been considered serious sexual deviance in earlier times (Peiss & Simmons, 1989, pp. 4-5).

As the 20th century churned on, the whiff of revolution was in the air. Sexuality became more open and pleasure oriented than it had ever been. People discussed and reflected on sexual matters incessantly and publicly while they reassured one another that sexuality was a secret, private, and largely insignificant pastime (Foucault, 1978, p. 35). Falling in lust became a respectable alternative to falling in love. Sexual relationships became valuable in their own right, even more so because they no longer had to involve reproduction and long-term commitments. With women's pursuit of sexual pleasure, encouraged by women's liberation, with more effective contraceptive technologies, and with a new feminine identity and sense of self came increasing tensions between men and women and the young and the old (Giddens, 1992). Problems surrounding teenage pregnancy were used as a way to condemn and to control the sexuality of adolescent girls because moral and medical forms of control were becoming less effective (Nathanson, 1991, pp. 163-164). People used sexuality as a way to challenge rules that they found oppressive and to achieve self-liberation and rewarding personal experiences (Quadagno & Fobes, 1995, p. 183).

With the arrival of what Janus and Janus (1993) called the "Second Sexual Revolution" in the early 1980s, the casual sex of previous years was replaced by something close to a sexual anarchy where practically anything went. The cornerstones were sexual activities among the mature and postmature generation and sexual activities that would have been considered quite deviant in earlier times (pp. 15-17). In a process of "defining deviancy down," people were prepared to accept sexual activities among others—or even to try some themselves—that would have been extremely shocking to the moral sensibilities of earlier times. The specter of AIDS and other sexually transmitted diseases was one restraint, but even the prospect of contracting a communicable disease was not enough for people to substantially alter their sexual activities. Sex was just too important to individuals for them to allow too much to interfere with their pursuit of sexual pleasures.

In the Eyes of the Beholder:
Attacking Autostimulation

Masturbation seems to be a regular sexual practice. In one study, 40% of females and approximately 60% of males (ages 18 to 59) reported that they had masturbated at least once during the previous year; about 25% of the male and 10% of the female respondents reported that they masturbated at least once a week (Michael et al., 1994, p. 158). The Janus report on human sexuality reported that 5% of the 1,338 males in their sample and 8% of the 1,398 females stated that their *preferred* way to achieve orgasm was through masturbation (Janus & Janus, 1993, p. 98). The people who are most likely to masturbate are individuals who have partners with whom they are sexually active. This suggests that individuals with the fullest sex life are the ones who are the most likely to masturbate (Michael et al., 1994, p. 165). One cornerstone of any successful moral crusade—and the masturbation mania of the 19th century certainly qualifies—is to disseminate successfully the belief that immoral actions of individuals have dire consequences for an entire society (Sherkat & Ellison, 1997, p. 974).

Both Judaism and Christianity condemn masturbation. The biblical inspiration for this denunciation is the sin of Onan. Rather than consummate a sexual relationship with his sister-in-law after the death of his brother, Onan spilled his semen on the ground, and the Lord struck him dead for his deed. It is not entirely clear if his sin was masturbation or his failure to honor the custom of levirate and impregnate his sister-in-law, but over the years the religious interpretation rigidified, and the cardinal sin of Onan came to be his act of autostimulation. Good Christians quickly learned that all righteous sexuality had to be oriented toward reproduction and then done with a minimum of energy and pleasure. Spending the seed without a good reproductive reason led to tragic results.

The condemnation of masturbation gained a stronger foothold as it went from being branded as badness to being branded as sickness. The prevailing and authoritative medical view of human afflictions during the 18th century was that supreme health came from a fine balance between bodily substances such as blood, bile, and sexual fluids. Any practice that upset this balance, the story went, would cause harmful results. Masturbation was especially suspect. Not only did it lack any apparent social purpose, but it depleted the body of essential fluids and generated far too much excitement in the masturbator (Michael et al., 1994, p. 159). Richard von Krafft-Ebing, a neurologist and psychiatrist born in Germany in the 19th century, concluded from his case studies that four categories of sexual deviation existed—masochism, sadism, fetishism, and homosexuality—all of which were abnormal and all of which were caused by masturbation (Von Krafft-Ebing, 1947). Male gynecologists were inclined to view women's sexual urges as quite unnecessary or even as dangerous distractions. The most troublesome of the lot was female masturbation. Once a female fell under its spell, all was lost, and her obsession with masturbation would know no bounds (Barker-Benfield, 1976, pp. 272-274). Even at the beginning of the 20th century, female masturbation was still seen as a major cause of women's frigidity in marriage (Jacquart & Thomasset, 1988).

The social reformers of the 19th century believed that the only way to halt the creeping social decay that they believed was plaguing U.S. society was for every individual to develop a strong personal moral code, especially with regard to food, drink, and sexual practices. A moral man or woman was an individual who was safe from all the temptation and corruption of city life because of his or her well-honed ability to recognize and then to control the ever-present dangers posed by the unholy trinity of sexual desire, stimulating food, and unhealthy practices. No challenge or temptation was so great that it could not be bested by a well-managed body. Sylvester Graham (1794-1851) was confident that total individual control was the one virtue that could make people truly free (Sokolow, 1983, p. 122). Graham's views were supported by others such as John Harvey Kellogg (1852-1943). Kellogg identified 39 warning signs of masturbation in a book that was published in 1888. Some of these were weakness, untrustworthiness, isolation, bashfulness, unnatural boldness, emaciation, paleness, colorless lips and gums, changes in disposition, underdeveloped breasts in females, gluttony, round shoulders, weak backs, pain in the limbs, stiffness of joints, lying on the stomach with the hands near the genitals, use of tobacco by boys, acne, biting the fingernails, obscene talk, and bed wetting (Money, 1985, pp. 91-98).

Both Graham and Kellogg had great faith in the power of correct food and proper eating habits to keep one morally upright. Masturbation was much less likely, they believed, if one learned proper nutrition at an early age. They insisted that masturbators were attracted to excessive amounts of spicy or upsetting foods such as pepper, salt, cloves, mustard, horseradish, or vinegar. Another telltale sign was the consumption of nonnutritious substances such as clay or plaster. Because they believed that nutritious or bland foods are strictly avoided by the chronic masturbator, Graham developed a mild foodstuff called the Graham cracker, and Kellogg developed the cornflake to help cure masturbation (Money, 1985).

If the dietary restrictions were unsuccessful, other things still might work. Parents were instructed to wrap up their children's genitals, to cover the genitals with some protective device, or simply to bind the masturbator's hands to make self-touching impossible. Devices were invented and peddled that supposedly could prevent male masturbation (Michael et al., 1994, p. 161). Most of these contraptions combatted masturbation by making it difficult, if not downright impossible, for a male to have an erection. One of these, a large ring with sharp spikes on its inside, was positioned to encircle a male's penis while he slept. If the penis were to become erect (tumescence), the spikes would come in contact with the organ, and pain, along with a rapid detumescence, would result. Another device was more elaborate and far less painful: It sounded an alarm when an erection occurred, alerting the vigilant, who would take the necessary steps to stop whatever was happening (Gregersen, 1983, p. 30). One treatment that was recommended to cure female masturbation was for the masturbator's parents to apply pure carbolic acid to her clitoris (Michael et al., 1994, p. 161).

It seems odd that masturbation (autostimulation or automanipulation) was once viewed as a most unacceptable practice and was said to have disastrous consequences for both the masturbator and society. Masturbation is now a practice that most people, if they think about it at all, define as relatively harmless, even though about half of the women and men who masturbate still report some guilt over it (Michael et al., 1994, p. 166). However, what about tomorrow? Could another Graham or Kellogg happen along and successfully attach new meanings to old practices?

Maligning Menstruation—
The Curse of the Cursed Curse

A biological event such as menstruation, found only among women of a particular age and status, is an event that can hardly be ignored, certainly not by a woman herself and certainly not by her peers. The variable meanings attached to this natural event remind us once again that social groups and social relationships must be at the center of our understandings of what is proper and improper. At some places and times, menstruation was a sign of an exalted status, and women were treated with veneration and respect. Menstrual blood was supposed to be able to cure leprosy, warts, hemorrhoids, epilepsy, and headaches. It was used as a love charm, a sacrifice to the gods, and a protection against evil spirits. A virgin's first menstrual pad was even saved and used as a curative for certain diseases (Delaney, Lupton, & Toth, 1988, p. 9). At other places and times, this periodic bleeding was viewed as a sign of contamination and pollution. Pliny, the Roman naturalist, thought that menstrual blood could ruin crops, dull steel knives, tarnish ivory, cause iron to rust, and make new wine turn sour. Dogs were supposed to be driven mad from the taste of menstrual blood, and bees supposedly died from it. Women were disallowed from working in the Saigon opium industry in the 1800s because opium was supposed to acquire a bitter taste if menstruating women were near it (Delaney et al., 1988, p. 9). At still other places and times, menstruation was viewed as an ordinary part of everyday life, and people did not worry very much about it.

Women's ability to produce blood that neither appreciably weakened nor killed them and that was associated with life and living rather than with death and dying probably produced a measure of envy among those men who were willing to think about it at all. Male initiation rites over large areas of Melanesia, Africa, the Americas, and Australia contain a rite of passage in which men actually emulate female menstruation by slitting the underside of their penises to facilitate bleeding. Sometimes the cut is short, and sometimes it extends the entire length of the shaft (Tannahill, 1992, p. 44). Why would men attempt to simulate female menstruation unless they viewed it as the source of a power that they wanted for themselves? (Knight, 1991, p. 37).

The way menstruation came to be symbolized—its social construction—reflected, at least in part, the conflicts and struggles between men and women for power and control. Laws (1990) argued that men tried to increase their dominance over women by constructing an ideology that defined women as naturally inferior. A cultural establishment of what menstruation meant and how it should be handled by women and responded to by men was a central part of this ideology (pp. 28-29). Times and places may be found where the menstrual taboos were exceedingly difficult for women to follow, and women most certainly did suffer a loss of status. They were secluded and isolated in special dwellings where they had little to do other than to handle their menstrual cramps as best they could and follow the rules that regulated only them (Mead, 1949, p. 222). At other times and places, however, females themselves were probably responsible for the custom of menstrual seclusion (Buckley & Gottlieb, 1988, p. 14). Perhaps the restrictions were not objectionable enough to fuss about, or perhaps women weighed the advantages and disadvantages of menstrual seclusion and found some benefit in acquiescing. Perhaps the restrictions imposed during the menstrual period were less objectionable than the restrictions on them at other points in their lives.

Even when the seclusion was not entirely voluntary, women could have turned it partly to their own advantage. It was usually a group event, and it provided opportunities for fellowship, relaxation, and spiritual growth (if the menstrual pains were not too great). It was a time for women to get in touch with other women while they got in touch with themselves and the rhythms of their own bodies (Hathaway, 1998). The segregation of menstruating women was an important event in the life of the entire community, and men's activities were strongly influenced by the menstrual cycles of women. Some men even had illicit love affairs with menstruating women during their period of isolation (Buckley, 1988).

To those women and men who consider menstruation as impure and debilitating comes a new view of this very natural process. The biologist Margie Profet (1993) argued that it is most unlikely that women would lose so much blood and be incapacitated for so long simply to slough off some unnecessary uterine lining. She stated that it is more likely that menstruation evolved as a way for a female to purge herself of infectious microorganisms (p. 338). Bacteria are carried from the woman's cervix or vagina into her uterus by getting on the heads or tails of swimming sperm, and seminal fluid can transport viruses from the male to the female reproductive organs (p. 341). Menstrual blood offers some protection to a female. It does help to remove uterine tissue that might be infected, but it also contains chemicals that kill a wide range of pathogens in the uterus, cervix, vagina, and oviducts (pp. 345-346). This view of menstruation is part of an emerging image of female sexuality that views females' sexual responses as having evolved to allow them to couple with whomever they liked, in rapid succession, without getting sick in the process from any intimate contact that they might have had with unclean or infected males.

> The composite picture is one of a female who seeks sexual pleasure and has developed biological defenses that protect her from the consequences of an active sex life, a creature designed by natural selection to improve her reproductive success by carefully directing her intimate life. (Small, 1995, p. 94)

The curse comes to an end by being defined as a blessing.

Summary and Conclusions

Whenever pigeonholes are created and people and what they do are forced into them, our ability to recognize the full range of human sexual diversity is restricted. Various possibilities arise from the interplay of biology, personality, culture, and society. Whenever differences are established—between individuals or groups—certain characteristics are ranked high and other characteristics are ranked low. When this asymmetry is coupled with intolerance and the power differentials between different groups, usually one category of individuals is defamed and devalued, and its members are secluded and isolated. Human sexuality, deviant and otherwise, is powerfully influenced by social and cultural factors. Because of this, sexual customs are relative, and the evaluations of sexual experiences and responses—some of which are physiological—are variable.

Fluctuations in the social meanings of both masturbation and menstruation are particularly instructive. They show us that any human event or practice can be the source of differentiation and evaluation. Depending on place, time, and situation, things sexual and things related to sex and gender are the object of reverence, scorn, or anything in between. The experiences of people who are stigmatized and isolated for their sexual outlooks and sexual interests are not appreciably different from the experiences of people who are scorned for other reasons.

References

Barker-Benfield, G. J. (1976). *The horrors of the half-known life: Male attitudes toward women and sexuality in nineteenth-century America.* New York: Harper & Row.

Beisel, N. (1993). Morals versus art: Censorship, the politics of interpretation, and the Victorian nude. *American Sociological Review, 58,* 145-162.

Birken, L. (1988). *Consuming desire: Sexual science and the emergence of a culture of abundance, 1871-1914.* New York: Cornell University Press.

Buckley, T. (1988). Menstruation and the power of Yurok women. In T. Buckley & A. Gottlieb (Eds.), *Blood magic: The anthropology of menstruation* (pp. 187-209). Berkeley: University of California Press.

Buckley, T., & Gottlieb, A. (1988). A critical appraisal of theories of menstrual symbolism. In T. Buckley & A. Gottlieb (Eds.), *Blood magic: The anthropology of menstruation* (pp. 3-50). Berkeley: University of California Press.

Bullough, V. (1976). *Sexual variance in society and history.* New York: John Wiley.

Burstyn, L. (1997). Female circumcision comes to America. In L. Salinger (Ed.), *Deviant behavior 97/98* (2nd ed., pp. 19-23). Guilford, CT: Dushkin.

Buss, D. (1994). *The evolution of desire: Strategies of human mating.* New York: Basic Books.

Byne, W. (1994, May). The biological evidence challenged. *Scientific American, 270,* 50-55.

Carter, S., & Sokol, J. (1989). *What really happens in bed: A demystification of sex.* New York: M. Evans.

Chodorow, N. (1994). *Femininities, masculinities, sexualities: Freud and beyond.* Lexington: University Press of Kentucky.

Delaney, J., Lupton, M. J., & Toth, E. (1988). *The curse: A cultural history of menstruation* (Rev. ed.). Urbana: University of Illinois Press.

Forsyth, C. (1996). The structuring of vicarious sex. *Deviant Behavior, 17,* 279-295.

Foucault, M. (1978). *The history of sexuality* (Vol. 1). New York: Pantheon.

Gagnon, J., & Simon, W. (1970). Perspectives on the social scene. In J. Gagnon & W. Simon (Eds.), *The sexual scene* (pp. 1-21). New Brunswick, NJ: Transaction.

Gagnon, J., & Simon, W. (1973). *Sexual conduct.* Chicago: Aldine.

Gay, P. (1986). *The bourgeois experience: Victoria to Freud: Vol. 2. The tender passion.* New York: Oxford University Press.

Giddens, A. (1992). *The transformation of intimacy: Sexuality, love and eroticism in modern societies.* Stanford, CA: Stanford University Press.

Goode, E., & Troiden, R. (Eds.). (1974). *Sexual deviance and sexual deviants.* New York: William Morrow.

Gregersen, E. (1983). *Sexual practices: The story of human sexuality.* New York: Franklin Watts.

Hathaway, N. (1998). Blood rites: Myths, taboos, and the eternal rhythm of the female body. In L. Salinger (Ed.), *Deviant behavior 98/99* (3rd ed., pp. 22-25). Guilford, CT: Dushkin.

Irvine, J. (1990). *Disorders of desire: Sex and gender in modern American sexology.* Philadelphia: Temple University Press.

Jacquart, D., & Thomasset, C. (1988). *Sexuality and medicine in the Middle Ages.* Cambridge, UK: Polity.

Janus, S., & Janus, C. (1993). *The Janus report on sexual behavior.* New York: John Wiley.

Kelly, G. (1994). *Sexuality today: The human perspective* (4th ed.). Guilford, CT: Dushkin.

Kemper, T. (1990). *Social structure and testosterone.* Brunswick, NJ: Rutgers University Press.

Kimmel, M. (1996). *Manhood in America: A cultural history.* New York: Free Press.

Kinsey, A. C., Pomeroy, W. B., & Martin, C. E. (1948). *Sexual behavior in the human male.* Philadelphia: W. B. Saunders.

Kinsey, A., Pomeroy, W. B., Martin, C. E., & Gebhard, P. H. (1953). *Sexual behavior in the human female.* Philadelphia: W. B. Saunders.

Knight, C. (1991). *Blood relations: Menstruation and the origins of culture.* New Haven, CT: Yale University Press.

Laumann, E. O., Gagnon, J. H., Michael, R. T., & Michaels, S. (1994). *The social organization of sexuality: Sexual practices in the United States.* Chicago: University of Chicago Press.

Laws, S. (1990). *Issues of blood: The politics of menstruation.* New York: Macmillan.

LeVay, S. (1993). *The sexual brain.* Cambridge, MA: MIT Press.

Lystra, K. (1989). *Searching the heart: Women, men, and romantic love in nineteenth-century America.* New York: Oxford University Press.

Mackie, G. (1996). Ending footbinding and infibulation: A convention account. *American Sociological Review, 61,* 999-1017.

Mead, M. (1949). *Male and female: A study of the sexes in a changing world.* New York: William Morrow.

Michael, R. T., Gagnon, J. H., Laumann, E. O., & Kolata, G. (1994). *Sex In America: A definitive study.* Boston: Little, Brown.

Money, J. (1985). *The destroying angel.* Buffalo, NY: Prometheus.

Money, J. (1988, February). The development of sexual orientation. *Harvard Medical School Mental Health Letter, 4,* 1-4.

Moon, D. (1995). Insult and inclusion: The term *fag hag* and gay male "community." *Social Forces, 74,* 487-510.

Moore, R. I. (1987). *The formation of a persecuting society: Power and deviance in western Europe, 950-1250.* New York: Basil Blackwell.

Nathanson, C. (1991). *Dangerous passage: The social control of sexuality in women's adolescence.* Philadelphia: Temple University Press.

Peiss, K., & Simmons, C. (1989). Passion and power: An introduction. In K. Peiss & C. Simmons (Eds.), *Passion and power: Sexuality in history.* Philadelphia: Temple University Press.

Profet, M. (1993, September). Menstruation as a defense against pathogens transported by sperm. *Quarterly Review of Biology, 68,* 335-386.

Quadagno, J., & Fobes, C. (1995). The welfare state and the cultural reproduction of gender: Making good girls and boys in the Job Corps. *Social Problems, 42,* 171-190.

Reiss, I. (1986). *Journey into sexuality.* Englewood Cliffs, NJ: Prentice Hall.

Ridley, M. (1993). *The Red Queen: Sex and the evolution of human nature.* New York: Penguin.

Rust, P. (1992). The politics of sexual identity: Sexual attraction and behavior among lesbian and bisexual women. *Social Problems, 39,* 366-386.

Saadawi, N. (1982). *The hidden face of Eve: Women in the Arab world.* Boston: Beacon.

Schur, E. (1988). *The Americanization of sex.* Philadelphia: Temple University Press.

Seidman, S. (1991). *Romantic longings: Love in America, 1830-1980.* New York: Routledge.

Sherkat, D., & Ellison, C. (1997). The cognitive structure of a moral crusade: Conservative Protestantism and opposition to pornography. *Social Forces, 75,* 957-982.

Small, M. (1995). *What's love got to do with it? The evolution of human mating.* New York: Anchor.

Sokolow, J. (1983). *Eros and modernization: Sylvester Graham, health reform, and the origins of Victorian sexuality in America.* Cranbury, NY: Associated University Presses.

Spain, D. (1992). *Gendered spaces.* Chapel Hill: University of North Carolina Press.

Tannahill, R. (1992). *Sex in history* (Rev., updated ed.). Chelsea, MI: Scarborough House.

Tavris, C. (1992). *The mismeasure of woman.* New York: Simon & Schuster.

Tewksbury, R. (1994). Gender construction and the female impersonator: The process of transforming "he" to "she." *Deviant Behavior, 15,* 27-43.

Tewksbury, R. (1996). Cruising for sex in public places: The structure and language of men's hidden, erotic worlds. *Deviant Behavior, 17,* 1-19.

Tiefer, L. (1978, July). The kiss. *Human Nature, 1,* 29-37.

Von Krafft-Ebing, R. (1947). *Psychopathia sexualis* (12th ed.). New York: Pioneer.

Weinberg, M., & Williams, C. (1994). *Dual attraction: Understanding bisexuality.* New York: Oxford University Press.

Weitz, R., & Bryant, K. (1997). The portrayal of homosexuality in abnormal psychology and sociology of deviance textbooks. *Deviant Behavior, 18,* 27-46.

Williams, W. (1986). *The spirit and the flesh: Sexual diversity in American Indian culture.* Boston: Beacon.

Ziv, L. (1997). The tragedy of female circumcision: One woman's story. In L. Salinger (Ed.), *Deviant behavior 97/98* (2nd ed., pp. 9-12). Guilford, CT: Dushkin.

4

The Relativity of
Predatory Violence

Introduction: The Concept of Murder

One of the inevitable facts of life is that people get injured and die, or at least their physical bodies do. Some of these deaths are caused by factors that seem to be beyond the immediate or direct control of human beings. These deaths are called accidental or natural. Some deaths, however, are deemed to result from deliberate human acts. These killings are considered to be malicious and unnecessary, and the individuals responsible for them can be subjected to punishments for what they have done. In the United States, illegal killings are called murders, and attempts to murder are called aggravated assaults.

The symbolic nature of human experience and the complexity in human social organization make it impossible to know for certain who or what is violent or how violence will be evaluated by different groups. No act is universally defined as violent and then negatively sanctioned in all human societies. In some places, for some people, certain acts of violence are viewed as normal and reasonable—as evidence of courage and bravery—and at other times, for other people, what appear to be similar acts are viewed as serious public health emergencies (Koop & Lundberg, 1992).

Violence Is in the Eyes and Actions of the Beholders

Unless the senses are sharp, it is easy to miss all the direct and indirect ways that beholders of some violent event actually create that which they believe they simply uncover. One thing that beholders do is to define violence into existence. It is the viewing audience that decides that some otherwise ambiguous event qualifies as a form of violence and then that it is abnormal or improper. Not all death and injury are necessarily viewed as violent, and not all violence is inevitably viewed as bad and unacceptable.

> A Roman holiday involved the slaughter of, literally, thousands in gladiatorial shows. After Spartacus' uprising, the crucified bodies of six thousand slaves lined the road from Rome to Capua like lampposts. In Christian Europe, executions replaced the Roman circuses. Criminals were beheaded publicly; they were hanged, their intestines drawn out and their bodies quartered; they were guillotined and elaborately tortured in front of festive crowds; their severed heads were exposed on pikes, their bodies hung in chains from gibbets. The public was amused, excited, more delighted than shocked. An execution was like a fun fair, and for the more spectacular occasions even apprentices got the day off. (Alvarez, 1972, pp. 53-54)

The culture-bound, historically specific definitions of violence do produce a great deal of selective perception and selective inattention. However, this does not exhaust the influence that the audience has. Secondary processes exist in which definitions and social reactions actually channel and change the course that violence takes.

Loseke (1992) showed how these secondary processes work with spouse abuse, but her analysis has relevance for any kind of violence. She was a participant observer at a shelter on the "South Coast" for battered women. She found that shelter personnel shared and then transmitted to others a particular worldview of spouse abuse and abused spouses. This worldview—that abuse is severe and ongoing and escalates over time and that battered women are so emotionally crippled by their victimizations that they have lost the will to leave the abuser—strongly influenced the day-to-day activities of agency personnel. Their core beliefs did not reflect an objective reality about predatory violence as much as they did a passionate desire of shelter personnel to garner public support for what they did. Subtly but persuasively, the message was sent out that shelters are beneficial for the victims of spouse abuse and that shelter personnel are knowledgeable and caring enough to make a difference.

A woman who appeared at the shelter for help but who did not fit the right profile—who appeared in control, was not the victim of the kind of abuse that shelter personnel felt comfortable to treat, or resisted the superficial interpretations of her experiences offered by shelter personnel—was simply sent packing. The experiences of victims of abuse were simplified, sometimes quite incorrectly and inappropriately, in the process of accepting and rejecting clients (Loseke, 1992, p. 54). Through their procedures, shelter personnel managed to perpetuate both the type of problem and the type of victim that the agency was prepared to handle (p. 114). The image of the world outside the shelter was made to fit the core beliefs and worldview that prevailed at the shelter, and not the other way around (p. 163).

Lloyd (1992) found that adult investigators of sexual child abuse used a variety of strategies to nudge children whom they interviewed in the direction of confirming that they had indeed been sexually abused by adults. The investigators used leading questions; they affirmed children's claims of abuse; they disconfirmed claims of nonabuse; they offered their own proabuse versions of what happened to see how children responded; and they treated children's agreements as much stronger than they actually were. None of this means that abuse did not occur. It may have and with tragic results. What this research shows is that people who have some reason to find violence and misery in the lives of others can usually find it. These investigators played a role in the construction of a social reality that they believed they played no role in at all.

Something potentially problematic can achieve the status of serious deviance only through legitimation and sustained attention from the mass media, consideration by researchers and writers, and the organized efforts of individuals who are determined to keep the issue alive and in public view (Jenness, 1995, p. 158). Unless strong advocates, organizational dynamics, power, and institutional resources all coalesce to get some troubling event authoritatively established as a social problem, then it may remain ambiguous, the object of competing claims, or part of some other broader problem and lacking its own unique identity (Unnithan, 1994, p. 71). This lack of sustained attention is the principal reason that the battered-husband problem has *not* received the same kind of attention as the battered-wife/woman problem.

> Professional and mass media attention to the issue of battered wives has been instrumental in its creation and continuation as an identified social problem. Along with social movement/organizational factors, this attention has been crucial to the construction of the social problem called "battered wives." The *lack* of these two factors has been of considerable importance to the failure of battered husbands. (Lucal, 1995, pp. 105-106)

Whatever media attention was directed to husband battering quickly died out once its novelty disappeared.

Certain claims or understandings attain the *status* of objective fact and then are used as self-evident truths to establish the certainty of other claims that are flimsier until they too seem to be objective facts. At bottom, violence is a social construction, and it cannot reasonably be understood in any other way. If an Ashanti woman of West Africa calls a man a fool, it is considered a violent act, and she can be killed for it. However, if an Ashanti child dies before puberty, he or she is given no funeral rites. The body is simply ditched in the community garbage dump. In that culture, the death of a child is no big deal, nor is the killing of an insulting woman; however, insulting a man is a very serious matter indeed (Service, 1963, pp. 376-378). People have different understandings about what qualifies as a violent act, and what is defined as violence in one place or time may be defined differently in another place or time. People from different cultures or groups may also disagree regarding proper levels of violence. In one setting, people may be negatively sanctioned for too much violence; in another setting, they may be sanctioned for too little violence; and in still another, they may be sanctioned for committing the wrong type of violence (Edgerton, 1976, p. 46).

Arbitrary Definitions

Homicide is the killing of one person by another. Some of these killings are illegal, in which case they are called murder (Lester, 1991, p. 1). Murder is defined in the Uniform Crime Reports (UCR) of the United States as the willful (non-negligent) killing of one human being by another (Federal Bureau of Investigation [FBI], 1998, p. 15). Some killings are excluded from the UCR program because they are not considered willful: suicides, accidental deaths, and traffic fatalities. If a person dies of a heart attack as the result of being robbed or witnessing a crime, it is not classified as a murder or a homicide either. However, even some killings that are willful are not called murder because they are considered justifiable. It is considered justifiable if a peace officer kills an individual in the line of duty, or if a private citizen kills an individual who is committing a serious crime (a felony). In 1997, the justifiable homicide total was 621: 353 killings of felons by law enforcement officers in the line of duty and 268 killings of felons during the commission of a felony by private citizens (FBI, 1998, p. 24).

A popular psychiatric diagnosis that is used to explain individual violence is *antisocial personality disorder* (Bogg, 1994). The ambiguousness of the term—it is called *psychopathy* or *sociopathy* as well as *antisocial personality*—is paralleled by the confusion over its meaning. Despite strenuous efforts to make the term more precise, it is still used as a catch-all to describe bad people and bad behaviors that cannot be condemned in other ways. It is a way to explain how seemingly normal people can do seemingly abnormal things: They have a mental disorder that has afflicted their moral development. The trouble is that the term is circular—sociopaths are labeled as sociopathic because they do sociopathic kinds of things. The unrestrained pursuit of self-interest—the principal sign that identifies antisocial people—is too general for the diagnosis to have much value. Egregious rule breaking or chronic incivility may be a social fact, provoked by life experiences in communities where violence and aggression are common-place (Green, 1993, p. 261). If people are not taught virtue, compassion, and altruism, it is unlikely that they will display them in their dealings with others.

Terms such as *willful, intended, accidental, justifiable, negligent,* and *anti-social* do not have clear and unambiguous meanings. Decisions about what qualifies as violence and who is violent are often unreliable, invalid, and biased. Wilson (1991) even claimed that a distinction can be drawn between "good" and "bad" murders. The distinction is based on the premise that groups of people distinguish those people who need or deserve to die (a "good" murder) and those who have every reason to live but are killed anyway (a "bad" murder). The blacker the villain, the more ghastly the killing, and the purer the victim, the greater the likelihood that the killing of the victim will be viewed as bad and the killing of the offender (if it occurs) as good (Wilson, 1991, p. 2).

Why Is Violence Relative?

Social Structural Variables

We must understand that some types of lethal or injurious acts in some places are both normal and normative: They occur frequently, are well integrated with other parts of society and culture, and are defined as legitimate and reasonable. This means that some people can kill or injure certain other people in certain ways without it being defined as deviant and without the killers or assaulters being labeled as deviants or negatively sanctioned. However, other kinds of violence, by other kinds of people, may be considered immoral, sick, or evil. Human aggression can occur in a variety of ways depending on the opportunities to express it and the sanctions and symbols applied to it (Goetting, 1995, p. xxii).

In a society with a great deal of diversity and heterogeneity (as is found in the United States), a number of subcultures exist. The wealthy may associate primarily with the wealthy, the poor primarily with the poor, and people from different ethnic and racial groups primarily with people who share a similar identity and social history. This diversity of subcultures produces a diversity of life experiences, opportunities, and outcomes. People from some subcultures consistently do better in achieving their dreams than people from other subcultures. They have more money, more control over their lives, and better jobs, and they generally feel good about what they do and what they are. These conditions are likely to promote peace and security. In other subcultures, however, there is misery and degradation, and people who live there are frustrated and angry much of the time (Blau & Blau, 1982).

Brutal conditions can breed brutality. Certain subcultures can be characterized by excessive violence because they bind people together in ways that stimulate conflicts or tensions from which interpersonal violence can result (Linsky, Bachman, & Strauss, 1995; Wolfgang & Ferracuti, 1967). When a tradition of neighborhood support for violence exists, along with the presence of actual groups that champion aggression, then violence is more likely (Felson, Liska, South, & McNulty, 1994, p. 170; Pinderhughes, 1993). These subcultures provide opportunities to observe, learn, and act out the kinds of violence that would shock, upset, or baffle individuals from different subcultures.

Most homicides and aggravated assaults in modern societies, Black (1983) informed us, are really attempts by individuals to maintain control of upsetting situations and to help themselves (p. 36). The victim and victimizer usually know one another, and the injuries or killings grow out of some argument or dispute. The perpetrator may view the violent act as a punishment (as when a child is spanked for misbehavior) or as a retribution (as when a spouse is killed for acts of infidelity or for being verbally or physically abusive). Most people justify violent acts, even extreme ones, if they are used to maintain social control or to defend one's intimates or oneself from harm (Heimer, 1997, p. 806). In this situation, violence is rational (precipitated by situationally appropriate motives such as defense, self-help, or social control) and positively sanctioned (supported or encouraged by other people in the group or by subcultural norms). It may violate legal norms in some technical sense, but it violates neither custom nor tradition (Black, 1993). Black reminded us that what may appear to be violent activities—irrational and meaningless—to an outsider may be very reasonable and meaningful to the people committing them (and perhaps to their associates, friends, or relatives).

Dueling is a good example of how interpersonal violence can become traditional and customary (Kiernan, 1988). At one time, duels were disorderly gatherings, with witnesses and spectators often joining in the fight. However, duels eventually became more ritualized and formalized. Each duelist selected someone to serve as a second, who was supposed to make sure that the rules were enforced and that the duel was fair. If swords were used, they had to be the same length. If pistols were used, they had to carry the same charge or be of the same caliber. Neither duelist was supposed to be at a disadvantage by, for example, having to face the sun or to shoot into the wind. The individual who was challenged had the choice of weapons. In France and England, the main weapon was the sword, whereas in the United States it was the pistol. In some places, a duel was ended at the drawing of first blood, but in others the dueling code required death. Though a large number of incidents, some quite petty, could precipitate a duel—personal insults, arguments over property, affronts to personal honor, or even cheating at cards—failing to respond to a challenge, no matter how ridiculous it was, meant disgrace and dishonor. Individuals who were arrested by police for dueling (it became a federal crime in the United States in 1877) almost always lacked remorse or shame because they were duty bound to duel if challenged.

The creation of rules and the application of sanctions to outsiders can actually prolong conflicts between subcultures and perpetuate continued deviation (Black, 1989). Usually, one group has the power to make rules and to determine sanctions for members of other groups:

> A certain group of people feels that one of their values—life, property, beauty of landscape, theological doctrine—is endangered by the behavior of others. If the group is politically influential, the value important, and the danger serious, the members of the group secure the enactment of a law and thus win the co-operation of the State in the effort to protect their value. (Sutherland, 1956, p. 103)

Rules that are invented by members of one group to serve their own interests and that are foisted on everybody else are likely to do more harm than good, especially if they are widely viewed as the reflection of the narrow self-interests of the few. Their enforcement is likely to generate more conflict, more resistance, and more opposition, not less (Diamond, 1971). People will not voluntarily remain quiet and passive for long in a society that is unfair and unequal—and that society will be characterized by high levels of violence and aggression (Currie, 1985). Some individuals may come to worry excessively over the prospects of riotous disorder, and any social spark may lead to a full-fledged moral panic (Adler, 1996).

Ideologies can make it easier for some people to get what they want by legitimating their violent activities as necessary and reasonable while branding the violence of other people as dangerous and unreasonable. Sometimes one particular group's "ideology of violence" will be internalized by a majority of people in the society; at other times, support for this ideology will be very limited. It is even possible for members of a group to be seduced by their own ideology and come to believe what it says without question. If an ideology of violence is firmly in place, the violent acts of some groups will appear more normal, more legitimate, more necessary, and less in need of correction or cure than other groups' injurious activities.

The social construction of violence usually involves the social construction of a target of aggression, an "other." The "other" is depersonalized and objectified, which helps neutralize any lingering moral inhibitions against violence that might still remain among the aggressors (Gamson, 1995). Successful creation and portrayal of the "other" as subhuman and deserving of pain and suffering may make it possible for rather ordinary people to commit acts of extraordinary evil (Katz, 1994). If the ideologies that legitimate violence toward an "other" are persuasive enough, individuals may commit acts of extreme cruelty with very little social prompting, even when they are given clear opportunities to renege without penalty or embarrassment (Browning, 1992, pp. 169-183).

The thugs were a group of violent criminals who once lived in India. They murdered and robbed their victims in honor of Kali, the Hindu goddess of destruction. Her followers believed that if they worshipped her and carried out her bidding, she would bestow great power and great riches on them (Hutton, 1981, p. 15). The weapons that they used in their murders were represented as parts of Kali herself: The pickaxe was imaged as one of her teeth; the knife was imaged as one of her ribs; and the cloth that thugs used to strangle their victims (called a *roomal*) was imaged as a piece of her garment. The thugs believed that their robberies and murders were not only moral but admirable (Hollick, 1840, p. 11).

The thugs were a sneaky bunch, and they took advantage of a victim's trust to complete a murder. The gang usually got advance information about who the travelers were and what of value they carried with them. The favorite victims of the thugs were native soldiers on leave, men of nobility and bearing, and treasure carriers (Ahmad, 1992, pp. 94-95). Sometimes children accompanied the thugs, partly so that they could learn the ropes and partly to help disguise the thugs' true mission. The thugs would journey along with their intended victim until the time was ripe to commit the killing. Usually a member of the gang would throw a rope or cloth around the neck of the victim and hold one end tightly. Another thug would hold the other end, and the rope or cloth would be crossed behind the victim's neck and drawn very tight. The two thugs would push the head of the victim forward to guarantee that suffocation occurred quickly. To make certain that the job was rightly done, the thugs would kick and hit the immobile victim to damage vital organs (Thornton 1837, pp. 6-8). Victims were sometimes dispatched with weapons other than a rope or cloth: A pickaxe was used to bludgeon and disembowel a victim, or a knife to stab and to cut. Sometimes the thugs would travel several days with their intended victim to allay any suspicions.

After a killing, the thugs quickly disposed of the body. A hole, three or four feet in depth, was dug. The corpse was dumped face downward in the grave, and then the mutilations began. Deep gashes were made on the body, and the limbs were pulled from the sockets and forced into unusual positions. Part of the reason for butchering the corpse was to make decomposition swifter and discovery less likely. In some cases, the corpse was treated in this irreverent way because the thugs were angry that the victim had so little for them to take. The hole was then covered over, and the thugs left the killing site. Sometimes the corpse was put in a sack and carried to a convenient burial site or was simply dumped into a well (Thornton 1837, pp. 9-10).

If the killings were done in the proper way, with due attention to omens, the thugs could expect success, blessings from Kali, and great financial rewards. However, if a killing was bungled, the thugs responsible could expect divine retribution from Kali herself. A pair of jackals crossing the road in front of the gang was considered a bad omen, but a single jackal was a good one. The ass was an exceptionally good omen, and so was the owl. If a thug sneezed at the beginning of an expedition, it was a sign of bad things to come. If the gang met a woman carrying a water pitcher on her head, the journey would be prosperous, and the killers would return safely home. It was even better if she was pregnant. However, if the pitcher was empty, the journey would be filled with misfortune and failure (Thornton 1837, pp. 80-89).

The thugs believed that if they carefully followed the wishes of Kali, they would earn supernatural points that could pay off in a subsequent life. This is one reason that they showed so little guilt or remorse over their violence. Murder was defined by them as inevitable and necessary, and thugs who did it in the proper way were defined by their associates as heroes. Their killings were acts of worship and reverence that would allow them one day to attain a state of grace and a close affinity with Kali. However, their murderous ways were not at all acceptable to the British, who began a drive in 1831 to end thuggee. However, what if the thugs had been more successful at spreading their religious beliefs and violent customs to others? What then?

Social Encounters

An important part of every social encounter, violence included, is the motives of the participants. Motives are sometimes viewed as inner factors that cause individuals to act in certain ways. "He hit his brother because he was angry" identifies anger as the motive for the assault. But motives are more than this. They are vocabularies or verbal statements used by people in a social encounter to explain, define, and interpret social action as it develops (Mills, 1940). They are provided in response to the question "Why did you do that?" (p. 907). People use motives to understand and sometimes to explain what they do; they also use motives to interpret and make sense of what others do. Motives are themselves part of a social situation, and they can determine how social relationships unfold and how they are viewed and evaluated by others (p. 904). Individuals learn a large number of possible motives as they internalize relevant parts of their social heritage. Just as people learn the proper ways of acting, they also learn a finite number of proper motives for acting. In some instances, individuals will abandon entirely what they plan to do, or they will choose a different course of action because they are unable to find appropriate motives for their planned activity or motives that others will accept. Individuals can disagree over which motives are operating in a particular situation, and they can be confused over what motivated others just as they can be unsure over their own motives (pp. 911-912). People may offer motives for their own actions that are disbelieved by others, and people may negotiate with each other until an appropriate motive is found that almost everybody will accept. A politician insists that he or she was forced to go above the written law and put the Constitution at risk for the good of the country. Others believe it was the politician's greed, arrogance, or stupidity—not the good of the country—that was the real motivating factor.

How does it help us to know that motives are social constructions that are situationally specific? Let us consider an example. Professional killers—individuals who are paid money to kill other people—have a vocabulary of motive that makes it possible for them to kill people without guilt or regret. They frame their murders in such a way that the violence appears routine and necessary and the victims appear deserving of whatever happens to them. "Routine kill of targets for money" is a motive that allows professional killers to disattend to the fact that they are ruthlessly killing real human beings; this motive helps them to maintain the degree of coldness, detachment, and daring required by their professional activities (Levi, 1989). The availability of motives within a culture or subculture to explain to members of a society why they need to be violent in specific situations is an important determinant of variations in violence.

Vocabularies of motive also have important consequences for witnesses or observers to some untoward or unsettling deviation. The motives that people offer to explain their actions are part of the social filter through which others define and judge them and decide what they have done. Violence that seems to be activated by proper or reasonable motives—motives that appear to others in the social encounter as legitimate reasons for acting violently—is viewed differently from violence that is perceived to be caused by unknown, improper, unreasonable, or bizarre motives. Self-defense is a motive for violence that is easier for most observers to accept than the claim that violence was used to serve the Goddess Kali. If the motives seem reasonable enough, the activity may not even be defined as a form of violence in the first place.

Another important part of most social encounters is the giving and receiving of *accounts*. An account is a verbal statement that an individual offers to others to excuse or to justify some untoward act if it is called into question (Lyman & Scott, 1989, pp. 112-132). An account is used by an individual to ward off the unwanted negative sanctions of others. "I'm sorry I'm acting so silly, but I'm just not myself today" is an account. If it is accepted by others, it is less likely that an individual will be sanctioned for having done silly things.

Accounts are interjected into social relationships to affect perceptions about the intent, aim, or motive of the deviating individual. If an individual successfully uses an account, he or she is able to convince others that what happened was really not so bad or that even if it was, it was unintentional and beyond control. Accounts are sometimes so standardized within specific cultures or subcultures that they are routinely expected when conduct departs too much from what is supposed to happen (Lyman & Scott, 1989, p. 112). For example, a police officer may handcuff an arrested suspect without offering any account. Handcuffing is part of the job and it is expected. However, if an arrestee is brought to the police station bruised and battered, some account will have to be offered by the arresting officer. He or she might offer the standardized account that the suspect was "resisting arrest." If this account is to be believed, it will convince others that the police officer did not really want to hurt the suspect but was forced to by the suspect's failure to comply with the officer's commands.

If individuals engage in routine, taken-for-granted behavior, an account is usually not expected by members of the group. They are doing what they are supposed to do, and their intent is clear. In the United States, for example, most people do not usually wonder why married people have children or why adults pay income taxes by April 15th. This is ordinary and expected behavior. However, if individuals commit some unsettling or unusual behavior, they may be expected to offer an account to clarify their purpose or intent. Thus, a father may claim it was necessary to punish his child with a severe beating because his child was being too naughty for the misbehavior to go unpunished. If the account is successful, then the father is seen as a concerned parent rather than a violent one.

Understanding that differences exist in people's abilities to account helps us to understand much of the relativity of violence. Some individuals, due to either their skill or their access to specific statuses, will be able to account for their violence in ways that others cannot. If professional killers were able to convince others that the individuals that they murdered were unworthy of concern, other people would view the killings as more proper and acceptable ("good" murders). Differences in accounts will produce great differences in how violent acts are defined, evaluated, and sanctioned, or even if they are defined as violence at all. Successful accounts make it possible for an individual to excuse or to justify what he or she has done.

Variable Violence

People in different societies or groups within those societies do understand the world differently, and people in one setting will encourage or overlook events and happenings that would irritate or appall people in different settings. This means that cultural variations exist in the killing and injury of human beings, and people hold divergent views about if or when they can hurt or kill one another.

The prevailing standards in the United States today with regard to when life begins and when it can be taken away, ambiguous as they are, are clear in one respect: It is totally unacceptable for anybody to intentionally kill or abuse an infant or child. Even the most adamant proabortionist believes that once a child is born, he or she is a person and has the right to live and grow, and antiabortionists would no doubt agree. The killing or abuse of an infant or child is usually viewed as inexcusable—a monstrous act by either a very sick or a very sadistic individual (Kantrowitz, 1998, p. 53). However, infanticide—killing of babies and small children—has not always been considered a particularly savage act.

Infanticide was once a fairly common practice (de Mause, 1974). Childhood was considered a brief period of development during which children were expected to acquire adult skills and capabilities as quickly as possible. They were widely viewed as ungrateful dependents or unwanted parasites (McCoy, 1992). Middle- or upper-class parents believed that having too many children interfered with their lifestyles and their pursuit of pleasure. Working- and lower-class families also had their reasons for not wanting too many children. A new child meant another mouth to feed and another person to care for in a family where resources were limited. Time spent in the birth and care of a new child pulled the mother away from important duties that usually contributed to the survival of the rest of the family. If the mother were to die in childbirth—a distinct possibility given the low level of medical knowledge and skill—an important and valuable family member would be eliminated. A strong likelihood existed that children would die early in life, so parents were disinclined to invest much time, energy, and emotional feeling in them. Illegitimate children, deformed children, babies born to parents who did not want them, and infants born to families who were too poor to afford them were regularly killed outright or simply allowed to die through abandonment or neglect. In Italy of the 19th century, women who were pregnant and unmarried were taken into custody by police and sent to homes where they delivered and then had their illegitimate children taken from them, never to be returned. If they were unable to pay the delivery fee, they were forced to remain in the institution and serve as wet nurses for other infants (Kertzer, 1993).

As the years passed and the conditions of social life improved, children become more valuable. By the late 1600s and early 1700s, childrearing practices were already starting to change throughout the world. The Puritans, for example, elevated the status of children to new heights and stressed the importance of the childhood years. They believed strongly in the value of the family, especially family cohesion and integration. Corporal punishments of children were really acceptable only if they were authorized by community leaders and done to strengthen the family (Pleck, 1987). By the late 1700s, more women than ever before had access to birth control techniques, and the existence of some choice with regard to pregnancy meant that children were coming to be defined more as rewards than as burdens. Brutal and violent childrearing techniques were being abandoned, and a new closeness existed between parents and children. Parents were more affectionate toward children and more indulgent and tolerant of them than they had ever been.

Improvement in economic conditions, new civil liberties for all people, healthier living conditions, and better medical care continued to change attitudes about children and childhood. Children came to be viewed more and more as persons to be protected from the predations of others, partly because children deserved full protections in their own right and partly because children needed to be properly nourished if they were to reach adulthood successfully and make their own contributions to society. Child labor laws were passed in most Western societies because the exploitation of children by adults was no longer considered acceptable. (Massachusetts passed the first state child labor law in the United States in 1836.) By the 20th century in the United States, children were generally viewed as economically useless but emotionally priceless. Their value was found in the pleasure that they brought to others and their capacity for learning and future upward mobility. Children had attained a sentimental value in the Western world unheard of and unimagined at other times and in other places (Zelizer, 1994).

The sentimental value of children found in the western world is not universal. In some places, at some times, infanticide has been viewed as logical and reasonable—as an expression of parental duty and responsibility. The Netsilik are an Eskimo group famous for their custom of infanticide. They are an isolated tribe of hunters living on the Arctic coast of northern North America. They hunt seals, caribou, and salmon with harpoons and spears and make clothing and build snow houses (igloos) heated by soapstone lamps. When they did practice infanticide, girl babies were usually killed more frequently than boy babies (Balikci, 1970/1989, p. 147).

The Netsilik considered female infanticide as one way to increase their chances of survival in a harsh environment. They were hunters and gatherers, so they could not grow crops to feed themselves or control the movements of the animals on which their survival depended. Sometimes little was available to gather or to hunt, prospects became very bleak, and starvation was imminent. Under these harsh conditions, infanticide was a rational but desperate act:

> Once when there was a famine Nagtok gave birth to a child, while people lay around about her dying of hunger. What did that child want here? How could it live, when its mother, who should give it life, was herself dried up and starving? So she strangled it and allowed it to freeze and later on ate it. (Rasmussen, quoted in Balikci, 1970/1989, p. 151)

Why kill girls instead of boys? The Netsilik defined females as both poor hunters and big eaters, so their loss from society was considered less objectionable than the loss of males. They believed that female infanticide increased the entire community's chances for survival by reducing the number of people who ate food but who were deemed less able to hunt for it (Balikci, 1970/1989, p. 151).

Female children were also killed under more normal conditions of life, when adults were neither angry and frustrated nor unusually hungry. Even here, however, infanticide can be understood as an outcome of Netsilik social organization and their desire to survive. Because men were defined as hunters and bringers of both food and life and women were not, this state of affairs elevated the status of male at the expense of the status of female. If a woman gave birth to a daughter and cared for her, it was considered time poorly spent; the mother could have given birth to a son and raised him in an equal amount of time. Thus, a daughter might be killed to make it possible for the family to produce a son.

Girl babies were killed in several ways. If a birth occurred during the winter, the newborn was placed at the entrance to the igloo. Her screams only lasted a few minutes because she quickly froze to death. In summer, a newborn was placed in a small grave, dug near the family's dwelling. The infant usually cried several hours before dying. A third method was used for all seasons: A furry skin was placed over the infant's face until the baby suffocated.

The decision to kill a child could be made by many people—the father, the mother, the grandfather, or the widowed grandmother. Patterns of authority in a Netsilik household were not rigid, and it was possible for an individual with a strong personality to have a significant impact. An older woman might speak with enough strength and conviction that she was able to determine the fate of a newborn. In the usual case of female infanticide, however, the biological father made the final decision, and the mother carried out the killing. The decision to kill a child was made as soon after birth as possible and without fanfare.

Infanticide was a flexible custom, not a rigid social rule, and alternatives to it existed. One important restraint on female infanticide was the act of naming the child. Children in the United States are considered persons at birth even though they may not yet have names. When they are given names, it is because parents like the way the names sound, because of fashion or family tradition, or because names identify them as unique or important individuals (Lieberson & Bell, 1992, p. 549). Dusin babies in North Borneo are not given formal names until they reach the age of 5. The Dusin believe it takes this long for an individual's personality to develop and that the child's name must reflect his or her unique qualities. Until the perfect name can be found, children are simply called by a nickname based on some habit or quirk that they may have, such as "holding on tightly" or "belly sticking out." In some regions of the world with high rates of infant mortality, parents may give their newborn a particularly unflattering name to dissuade unfriendly spirits from first claiming and then killing their child (Duham, Myers, Barnden, McDougall, & Kelly, 1991, pp. 158-159).

The Netsilik believed that the naming of a child was very spiritual and that names had supernatural power (Balikci, 1970/1989, p. 148). Births were viewed as events in which spirits of the dead were present that watched the proceedings with a great deal of interest. The spirits were unable to speak, but they could still have a strong impact on the birthing process. During a particularly difficult part of a child's birth, a woman might cry out names of spirits to garner help from one of them. If one name seemed more helpful in the delivery than all others, this name was given to the newborn. For example, one female child among the Netsilik was named Manelaq because when that name was shouted out by her mother, the child was born (Balikci, 1970/1989, p. 149).

Mothers tried to find a name quickly in the hopes of enlisting the help of some concerned spirit that would make childbirth easier and less physically painful. With the name came an identity and character for the newborn. It also brought a patron spirit that now had a new home in the child. Parents were much less likely to kill a named child because it might offend the spirit who occupied the newborn's body, and offending spirits is something that groups with a precarious existence try to avoid if at all possible. Sometimes the child could not be named because no name seemed to fit or because the naming ritual simply did not occur at the time of the child's birth. An unnamed child could be more easily killed. Though naming of a newborn did seem to help parents with the birth process, it also helped to keep some infants alive.

The Netsilik were either unconcerned with or unaware of the grave dangers that female infanticide posed to them. Beyond the loss of infant life, female infanticide produced powerful tensions and a great deal of disharmony. It produced an imbalance in the sex ratio and reduced the number of marriageable women. This increased the pressure on men to find and hold eligible mates. This competition among men divided the community into many small, mutually suspicious groups. In some cases, husbands were murdered—they were simply stabbed in the back by envious men who wanted to marry their wives. The community had to find some balance between the need for females' reproductive contributions and the need to maintain food supplies at subsistence levels.

In 1936, a Catholic missionary established a mission near the Netsilik. As a result of his preaching and ongoing personal involvement with them, he was able to suppress the practice of female infanticide. This eventually caused a much greater balance in the sex ratio between men and women. Men were able to find wives much more easily, and they were much less likely to murder each other to obtain mates. Monogamy became the rule, and other marital arrangements (such as polygyny or polyandry) fell into disfavor. Very quickly, according to Balikci (1970/1989), the Netsilik started acting like proper Christians (p. 250).

In some places, killing a person who is legally determined to "deserve it" because of some injury he or she has visited on the community receives great support. The killing—called capital punishment or the death penalty—may not even be viewed as a violent act because of its legitimacy in the eyes of the beholders. Unlike dueling violence, which tended to skirt the official apparatus of control, the deaths produced by capital punishment occur according to strict rules and formalized procedures carried out by representatives of the government itself. Specific due-process requirements must be followed for an individual to be sentenced to death and then executed, and the convict is able to appeal the sentence, many times in some cases.

Not all states in the United States have the death penalty, and even within the same community or family, people can hold different opinions about its propriety (Wekesser, 1991). Attitudes toward capital punishment in the United States are volatile, having shifted during the last 30 years or so from a state of weak support for capital punishment (42% favored it as a punishment for murder in 1966) to the strong support of the 1980s and 1990s (71% favored it as a punishment for murder in 1996) (Maguire & Pastore, 1998, p. 141).

Ancient executions were savage affairs, characterized by a great deal of rage and anger, in which the offender was expected to die slowly and in great misery. All members of the community participated, either as executioners or as witnesses (Johnson, 1990, p. 6). The bodily organs of some offenders were mutilated to punish their owners even more. Intestines, hearts, and kidneys have been ripped out of the bodies of condemned individuals, sometimes while they were still alive, and hands, noses, tongues, and ears have been either cut off or mutilated. In some cases, this was done because some body parts—intestines, hearts, kidneys, hands—were viewed as the organs responsible for crimes. Sometimes offenders were mutilated simply to prolong their suffering or to humiliate them more (Newman, 1978, pp. 47-50).

The identifying features of modern executions are their bureaucratic nature and their isolation from the community (Johnson, 1990, pp. 18-31). Community feelings of rage are supposed to be absent or at least suppressed. Over the years, public executions have become much less public. What was once an act by a community on its own behalf has become the special province of prison officials who are expected to approach executions with professionalism and detachment. The last government-sponsored public execution in the United States occurred in Owensboro, Kentucky, where a young black man was hanged for assaulting a 70-year-old woman. Twenty thousand people attended the execution, and they were a coarse, unruly crowd (Prejean, 1989, p. 101).

No doubt exists that some individuals are prepared to use brutality and violence to get what they want and that they produce untold misery for everybody else. Consider the following account written by Mrs. Kuzmaak, describing the anguish caused by the murder of her daughter, Donna.

> I keep searching for the right descriptive words to convey the emotional impact our daughter's murder made on us. "Devastating" just doesn't make it—"ravaged" is closer, engulfed, overwhelmed, drowning in sadness, numb, oblivious to EVERY-THING else, totally immersed in the horror, the why, the who, what she had to go through in the closing minutes of her short life, how terrified she must have been, did she scream for help and no one came, did she fight, the pain, how it felt to be strangled, what her dying thoughts were, how she must have held out hope until the last that she would be rescued, her shock and disbelief that this was happening, and as the information unfolded itself to us in bits and pieces that first day, the anguish of hearing how badly she was beaten, then a couple of hours later crying out when I heard that she had been repeatedly stabbed. Then, the ultimate horror to learn that the cause of death was strangulation. To be deprived of breath—lungs bursting—"Oh, God, oh God," I would wail, tears streaming, hands clenched and imploring.
>
> That is how I remember the day, March 21, 1979. A decade has passed, but the emotions go on, the anger, the sorrow, and the loss. (Hickey, 1997, pp. 125-126)

Murderers not only take lives needlessly and ravage the families and friends of the victim but also violate the sentiments of trust and affiliation that serve as the foundation of social life. The anger, the hurt, and the desire for revenge are easy enough to understand (Berns, 1982, pp. 334-335). However, not everyone who has had a family member savagely murdered is in favor of the death penalty:

> I was eight years old when my father was murdered. It is almost impossible to describe the pain of losing a parent to a senseless murder. And in the aftermath, it is similarly impossible to quiet the confusion: "Why him? Why this? Why me?" But even as a child one thing was clear to me: I didn't want the killer, in turn, to be killed . . . I saw nothing that could be accomplished in the loss of one life being answered with the loss of another. And I knew, far too vividly, the anguish that would spread through another family—another set of parents, children, brothers, and sisters thrown into grief. (Kennedy, 1989, p. 1)

The author of this is M. Kerry Kennedy, the daughter of Robert Kennedy. Her father was brutally gunned down in Los Angeles in June 1968 during his bid for the Democratic nomination for the presidency of the United States. Why did Ms. Kennedy believe so strongly in compassion and nonviolence, even as a child and after the murder of her father, whereas some other people whose lives have not been as dramatically changed either will not or cannot speak out against the death penalty? Do they have greater concern for the victims of violence than Kennedy, or is something else operating?

At the heart of the controversy over capital punishment is a more subtle controversy over the meaning of state-sponsored killing. It is easier for most people to view executions as less violent than the killing that the offender committed that got him or her sent to the death chamber in the first place. The heightened fear of violent crime, coupled with excessive media coverage of mass murders and senseless killings, no doubt plays a part in people's growing support for capital punishment (Rankin, 1979, p. 197). However, more is at work than this. Executions are carried out by that segment of a society that is legitimately empowered to enforce criminal law and to administer criminal sanctions. This segment has the resources, not only to carry out an execution but to present it as necessary and justified. Executions appear to be activated by proper motives such as retribution and protection of innocent victims, and executioners are able to offer appropriate motives and credible accounts for their conduct. It is widely believed that individuals who cannot live peacefully in an open, free, and prosperous society such as the United States deserve what they get, even if it is the death penalty.

Elements can be found in an execution of a convicted felon to suggest that it is—or could be interpreted as—a premeditated violent act. Johnson describes an execution he witnessed of a felon named Jones. Jones was strapped into the electric chair, the electrodes were attached to his head and leg, and a tight mask was placed over his head. Only his nose was visible. At the designated time, the electric switch was thrown, and 2,500 volts of electricity, at 5 to 7 amps, passed through Jones's body, starting at his head and ending at the electrode on his ankle. The following describes the grisly details:

> Jones sat perfectly still for what seemed an eternity but was in fact no more than thirty seconds. Finally, the electricity hit him. His body stiffened spasmodically, though only briefly. A thin swirl of smoke trailed away from his head, then dissipated quickly. (People outside the witness room could hear crackling and burning; a faint smell of burned flesh lingered in the air, mildly nauseating some people.) The body remained taut, with the right foot raised slightly at the heel, seemingly frozen there. A brief pause, then another minute of shock. When it was over, the body was inert. (Johnson, 1990, p. 111)

The similarities between the murder *by* Jones (or people like him) and the execution *of* Jones (or people like him) may be greater than most people like to think.

With the crime of murder, a harm exists (a death), intent to kill (*mens rea*), and the deliberate taking by an offender of an individual's life. Likewise, an execution is characterized by harm (death of the convicted felon), intent to kill (the execution is meticulously planned and carefully rehearsed to ensure a smooth and successful killing), and the deliberate taking of an individual's life (execution of a convicted felon). In this regard, the crime of murder and capital punishment are identical (Kennedy, 1976, pp. 59-62). Mark Kennedy (1976) insisted that it is one group's superior power that creates and sustains the fallacy that punishment and crime are different from one another and belong to independent, mutually exclusive categories (p. 62). He stated that it is peculiar that some groups will go to great lengths to try to stop the violence of murder but will allow, or even encourage, the violence of capital punishment.

In the United States, an individual who kills an innocent victim—such as Donna Kuzmaak or Robert Kennedy—is a person whose "just desserts" might very well include an execution. However, as our exploration of infanticide shows, killing an innocent is not at all times and places a basis for an execution. Lesser (1993, p. 6) insisted that the unique qualities of a victim—how innocent, pure, or kind he or she is considered to be—cannot reasonably be used to determine a killer's level of criminal responsibility or what his or her penalty should be. "Just desserts" is itself a social construction and is subject to great variation from situation to situation. Viewing the meaning of the death penalty from the standpoint of *just* the victim (and the victim's family and friends) or *just* the accused (and the accused's family and friends) will make it impossible to appreciate the wider implications. The death penalty is a public issue, and its usage affects countless individuals, their families, and the entire social fabric (Turnbull, 1989, p. 168).

Summary and Conclusions

Violence—or social acts called violence—is a regular feature of most societies, and much of it is done by ordinary people who view their use of violence as practical and necessary, a form of self-help when legal authorities either will not or cannot help. These individuals seek their own brand of justice, and they seem to care little that they might be arrested and imprisoned for what they have done. Some of the individuals who commit murder may wait for police to arrive or may even report their own crimes to police. Regardless of what the law demands, these people do what they think is right, and they are prepared to suffer the consequences. Important factors in the variability of violence are the ability to offer appropriate motives for what one has done and to offer accounts if one's violent conduct is called into question.

Infanticide, capital punishment, dueling, and thuggee are not simply random events precipitated by stress and frustration. They almost always reflect patterns of social organization. What is the difference between the deaths caused by capital punishment, dueling, thuggee, or infanticide and other violent happenings in some society? Are there any objective, intrinsic, or inherent qualities of killings *themselves* that would allow a dispassionate observer to know which deaths are legitimate and which are not?

References

Adler, J. (1996). The making of a moral panic in the 19th-century America: The Boston garroting hysteria of 1865. *Deviant Behavior, 17,* 259-278.

Ahmad, I. (1992). *Thugs, dacoits, and the world-system in nineteenth century India.* Unpublished doctoral dissertation, State University of New York at Binghamton.

Alvarez, A. (1972). *The savage god: A study of suicide.* New York: Random House.

Balikci, A. (1989). *The Netsilik Eskimo.* Prospect Heights, IL: Waveland. (Original work published 1970)

Berns, W. (1982). The morality of anger. In H. A. Bedau (Ed.), *The death penalty in America* (3rd ed., pp. 331-341). New York: Oxford University Press.

Black, D. (1983). Crime as social control. *American Sociological Review, 48,* 34-45.

Black, D. (1989). *Sociological justice.* New York: Oxford University Press.

Black, D. (1993). *The social structure of right and wrong.* San Diego, CA: Academic Press.

Blau, J., & Blau, P. (1982). The cost of inequality: Metropolitan structure and violent crime. *American Sociological Review, 47,* 114-129.

Bogg, R. (1994). Psychopathic behavior as perpetual gaming: A synthesis of forensic accounts. *Deviant Behavior, 15,* 357-374.

Browning, C. (1992). *The path to genocide: Essays on launching the final solution.* New York: Cambridge University Press.

Currie, E. (1985). *Confronting crime: An American challenge.* New York: Pantheon.

de Mause, L. (1974). *The history of childhood.* New York: Psychohistory Press.

Diamond, S. (1971). The rule of law versus the order of custom. *Social Research, 38,* 42-72.

Duham, C., Myers, F., Barnden, N., McDougall, A., & Kelly, T. (1991). *Mamatoto: A celebration of birth.* New York: Penguin.

Edgerton, R. (1976). *Deviance: A cross-cultural perspective.* Menlo Park, CA: Cummings.

Federal Bureau of Investigation. (1998). *Crime in the United States 1997.* Washington, DC: Government Printing Office.

Felson, R., Liska, A., South, S., & McNulty, T. (1994). The subculture of violence and delinquency: Individual vs. school context effects. *Social Forces, 73,* 155-173.

Gamson, W. (1995). Hiroshima, the Holocaust, and the politics of exclusion: 1994 presidential address. *American Sociological Review, 60,* 1-20.

Goetting, A. (1995). *Homicide in families and other special populations.* New York: Springer.

Green, E. (1993). *The intent to kill: Making sense of murder.* Baltimore: Clevedon.

Heimer, K. (1997). Socioeconomic status, subcultural definitions, and violent delinquency. *Social Forces, 75,* 799-833.

Hickey, E. (1997). *Serial murderers and their victims* (2nd ed.). Belmont, CA: Wadsworth.

Hollick, F. (1840). *Murder made moral.* Manchester, UK: A. Heywood.

Hutton, J. (1981). *Thugs and dacoits of India.* New Delhi: Gian.

Jenness, V. (1995). Social movement growth, domain expansion, and framing processes: The gay/lesbian movement and violence against gays and lesbians as a social problem. *Social Problems, 42,* 145-170.

Johnson, R. (1990). *Death work: A study of the modern execution process.* Pacific Grove, CA: Brooks/Cole.

Kantrowitz, B. (1998). Cradles to coffins. In L. Salinger (Ed.), *Deviant behavior 98/99* (3rd ed., pp. 53-54). Guilford, CT: Dushkin.

Katz, F. (1994). *Ordinary people and extraordinary evil: A report on the beguilings of evil.* Albany: State University of New York Press.

Kennedy, M. (1976). Beyond incrimination: Some neglected facets of the theory of punishment. In W. Chambliss & M. Mankoff (Eds.), *Whose law? What order?* (pp. 34-65). New York: John Wiley.

Kennedy, M. K. (1989). Foreword. In I. Gray & M. Stanley (Eds.), *A punishment in search of a crime: Americans speak out against the death penalty* (pp. 1-3). New York: Avon.

Kertzer, D. (1993). *Sacrificed for honor: Italian infant abandonment and the politics of reproductive control.* Boston: Beacon.

Kiernan, V. G. (1988). *The duel in European history: Honor and the reign of aristocracy.* New York: Oxford University Press.

Koop, C. E., & Lundberg, G. D. (1992). Violence in America: A public health emergency. *Journal of the American Medical Association, 267,* 3075-3076.

Lesser, W. (1993). *Pictures at an execution.* Cambridge, MA: Harvard University Press.

Lester, D. (1991). *Questions and answers about murder.* Philadelphia: Charles.

Levi, K. (1989). Becoming a hit man: Neutralization in a very deviant career. In D. Kelly (Ed.), *Deviant behavior* (3rd ed., pp. 447-458). New York: St. Martin's.

Lieberson, S., & Bell, E. (1992). Children's first names: An empirical study of social taste. *American Journal of Sociology, 98,* 511-554.

Linsky, A., Bachman, R., & Strauss, M. (1995). *Stress, culture, and aggression.* New Haven, CT: Yale University Press.

Lloyd, R. (1992). Negotiating child sexual abuse: The interactional character of investigative practices. *Social Problems, 39,* 109-124.

Loseke, D. (1992). *The battered woman and shelters: The social construction of wife abuse.* Albany: State University of New York.

Lucal, B. (1995). The problem with "battered husbands." *Deviant Behavior, 16,* 95-112.

Lyman, S., & Scott, M. B. (1989). *A sociology of the absurd* (2nd ed.). New York: General Hall.

Maguire, K., & Pastore, A. (Eds.). (1998). *Sourcebook of criminal justice statistics 1997.* Washington, DC: Government Printing Office.

McCoy, E. (1992). Childhood through the ages. In K. Finsterbusch (Ed.), *Sociology 92/93* (21st ed., pp. 46-49). Guilford, CT: Dushkin.

Mills, C. W. (1940). Situated actions and vocabularies of motive. *American Sociological Review, 5,* 904-913.

Newman, G. (1978). *The punishment response.* Philadelphia: J. B. Lippincott.

Pinderhughes, H. (1993). The anatomy of racially motivated violence in New York City: A case study of youth in southern Brooklyn. *Social Problems, 40,* 478-492.

Pleck, E. (1987). *Domestic tyranny: The making of American social policy against family violence from colonial times to the present.* New York: Oxford University Press.

Prejean, Sister H. (1989). A pilgrim's progress. In I. Gray & M. Stanley (Eds.), *A punishment in search of a crime: Americans speak out against the death penalty* (pp. 93-101). New York: Avon.

Rankin, J. (1979). Changing attitudes toward capital punishment. *Social Forces, 58,* 194-211.

Service, E. (1963). *Profiles in ethnology.* New York: Harper & Row.

Sutherland, E. (1956). Crime and the conflict process. In A. Cohen, A. Lindesmith, & K. Schuessler (Eds.), *The Sutherland papers* (pp. 99-111). Bloomington: Indiana University Press.

Thornton, E. (1837). *Illustrations of the history and practices of the thugs.* London: W. H. Allen.

Turnbull, C. M. (1989). The death penalty and anthropology. In M. L. Radelet (Ed.), *Facing the death penalty: Essays on a cruel and unusual punishment* (pp. 156-168). Philadelphia: Temple University Press.

Unnithan, N. P. (1994). Children as victims of homicide: Making claims, formulating categories, and constructing social problems. *Deviant Behavior, 15,* 63-83.

Wekesser, C. (Ed.). (1991). *The death penalty: Opposing viewpoints.* San Diego, CA: Greenhaven.

Wilson, W. (1991). *Good murders and bad murders.* Lanham, MD: University Press of America.

Wolfgang, M., & Ferracuti, F. (1967). *The subculture of violence.* London: Tavistock.

Zelizer, V. (1994). *Pricing the priceless child: The changing social value of children.* Princeton, NJ: Princeton University Press.

5

The Relativity of Sexual Violence

Introduction: Ideal Culture and Real Rape

Some individuals enhance their own feelings of power and control, release feelings of anger, and satisfy sadistic urges by persuading or actually forcing other people to engage in unwanted acts of sex (Groth, 1979). Few nonhuman animals can accomplish a mating without the cooperation of *both* partners (Gregersen, 1983, p. 54), but both the human female and the human male are susceptible to rape. Some individuals are both willing and able to consummate sexual acts with other individuals against their will through the use of force, persuasion, or trickery (Pelka, 1995, pp. 251-252).

Rape is the term usually used to refer to acts of sexual violence or coercive sexuality. Sometimes the word *rape* is used in a limited way to cover only enforced coitus between a male offender (or offenders) and a female victim (or victims), but at other times it is used to refer to a broader range of violent sexual behaviors such as forced oral sex, forced anal penetration, forcible fondling, or object penetration between all kinds of people in all kinds of situations. In this broader sense, not only can men rape women, but men can rape men, women can rape women, and women can rape men (Macdonald, 1995, pp. 13-19). Sexual harassment is even viewed by a growing number of individuals as a type of coercive sexuality because it, like rape, objectifies and depersonalizes victims, while it controls and injures them (Fitzgerald, 1993, p. 1072; McKinney, 1994).

Rapes vary in the motivations of the rapists, the act itself, the relationships between victim and rapist, and the personal characteristics of the offender and the offended (Allison & Wrightsman, 1993, p. 3). Disputes exist over the meaning of rape and over what kinds of sexual acts actually qualify as coerced (Gilmartin, 1994, p. 16). Rape is not a set of concrete actions that can be objectively and uniformly identified no matter what. It is a definition—a construction or interpretation, if you will—that varies from one observer, audience, or party to another; from one subculture and society to another; and from one historical time period to another. In short, different audiences define rape in different ways (Goode, 1996).

The Federal Bureau of Investigation (FBI) defines forcible rape as the carnal knowledge (vaginal intercourse) of a female forcibly and against her will (FBI, 1998, p. 25). Though this definition seems clear and direct, it is not applied clearly and directly. In actuality, legal authorities think of rape in a more restrictive and selective way: "It is deemed a rape only if the assailant is a violent stranger, if the victim reports the rape immediately after it occurred, and if she can provide evidence of the attack and of her active resistance" (Weis & Borges, 1973, p. 71). The determination that rape has occurred is really an outgrowth of the give and take between individuals who have different interests, resources, and understandings about what really happened. Consider the fact that the crime of forcible rape is "unfounded" by police more often than any other serious crime (FBI, 1998, p. 26). For a rape to be unfounded means that a police officer decides that no rape has occurred and that the victim's complaint is false, incorrect, or that insufficient grounds exist for a suspect to be arrested. The relatively high unfounding rate with rape (8% for rape as compared to 2% for other serious crimes) suggests that disagreement exists between police and female victims of rape about what rape really is and whether one has actually occurred.

Some states have expanded the meaning of *rape* to include acts in addition to vaginal penetration (e.g., forced oral sex; object penetration). They have replaced the term *rape* with terms such as *sexual assault, sexual battery,* or *criminal sexual conduct* to emphasize that sexual violence is violence, not sex. They have removed requirements that a victim must corroborate her charges of rape with some objective proof before she is believed. They have removed requirements that a woman must offer the utmost resistance to her attacker and be able to prove that she fought valiantly. They have adopted rape shield laws that have made evidence of the woman's past sexual behavior inadmissible in court. Rape law reforms are in a constant state of flux, and considerable variation exists from jurisdiction to jurisdiction in the United States (Epstein & Langenbahn, 1994, p. 7). Some states have moved quickly to make extensive reforms, but others have moved slowly and sluggishly (Spohn & Horney, 1992, pp. 159-175).

Forcible rape is no longer a capital crime in the United States. The U.S. Supreme Court decided in 1977 in *Coker v. Georgia* (433 U.S. 485) that the raping of an adult female was not serious enough to warrant capital punishment— that is, that capital punishment was grossly disproportionate and an excessive penalty for forcible rape (Bedau, 1982, p. 299). Think how the 455 men who have been executed for rape in the United States (Barlow, 1996, p. 395) would feel if they learned that it was not really what they did, who they were, or the region of the country in which they raped that got them killed; it was mainly *when* they did it. And think how rape victims must *now* feel. Many—perhaps most—rape survivors view rape as much more than an assault on their bodies. They view it as the death of the self (Carosella, 1995, p. xx). These women must find it incomprehensible that an experience so harmful to them is no longer considered serious enough to warrant the ultimate sanction.

Sexual violence is a principal source of anxiety, fear, injury, and sometimes even death for women, especially those who become direct victims of it (Ferraro, 1996, p. 675; Goodman, Koss, Fitzgerald, Russo, & Keita, 1993, p. 1054; Koss, 1993, p. 1062). This "female fear" is a special burden that women carry that is not shared with men who have not themselves been raped. Some of the fear exists because women believe—incorrectly, as it turns out—that a substantial number of rape victims are murdered by their assailants (Gordon & Riger, 1989). In those murders in which circumstances were known (between 1976 and 1994), 1.5% involved rape or some other sex offense (Greenfeld, 1997, p. 3). However, rape does exert a "shadow effect" in that women, especially young ones, are apprehensive about *any* victimization because they always fear that it could lead to a rape (Ferraro, 1996, p. 669).

The legal changes, if they accomplished nothing else, highlighted the need for reform. This publicized the ongoing disputes over the nature of rape and what is true and what is false about it. A great deal of attention was directed to rape and its victims, and rape is now an important issue in U.S. society.

> Only by an outmoded and unrealistic standard could the effort of the rape law reform movement be considered a failure. In short, there is evidence that public opinion on the subject of rape is changing and that victims' voices have been heard. (Cuklanz, 1996, p. 116)

The plight of the victim, the violence of rape, the cruelty of the rapist, and the sexism of both society and the criminal justice system—these are all now regular features of public discourse about rape and cultural representations of it.

The Symbolic Organization of Sexual Violence

The kinds of acts that are classified as sexual violence and how they are interpreted are highly variable, and the determination that sexual violence has occurred is conditioned by a large number of social and cultural factors (Palmer, 1989). No society has yet developed a purely objective way to measure the amount of force or the degree of consent that is present during an act of sex, nor has any society developed a purely objective way to determine at what point sex becomes coercive enough that it qualifies as rape. "The act must be interpreted as rape by the female actee, and her interpretation must be similarly evaluated by a number of officials and agencies before the official designation of 'rape' can be legitimately applied" (Svalastoga, 1962, p. 48). Cross-cultural comparisons of rape rates are practically useless, but even within the same society or community, the study of rape is filled with difficulties and complications (Chappell, 1976, p. 296). The outlooks of the victimizers may be, and probably are, different from those of the victimized (Fried, 1997, pp. 40-41).

Bourque (1989) interviewed a sample of 126 white and 125 black Los Angeles County residents during the summer and fall of 1979 to determine how they defined rape (p. 61). Her respondents used information about the assailant's degree of force and the victim's degree of resistance to decide if a rape had occurred, but they did not all reach the same conclusions. Some respondents focused on what the female victim did and were prone to blame her for what happened. They were inclined to believe that rape could be avoided if a woman resisted strenuously enough and that if rape occurred, it was because she did something to cause it. These respondents tended to view rape in sexual terms. Other respondents, by contrast, looked primarily at the assailant. If he used any force to achieve his sexual goals, they believed that a rape had indeed occurred. These respondents tended to view rape in violent terms. The rest of Bourque's sample did use information about force and resistance but also used details about location, acquaintanceship, ethnicity, prior sexual history, and socioeconomic status of the participants in deciding what had really happened (p. 204). Bourque found a great deal of variation in definitions of rape. White females were the most likely to use information on the degree of force as their principal cue and were the ones most likely to define a sexual encounter as rape; black males were the most likely to use information on resistance, and they were the least likely to label sexual encounters as rape (pp. 228-229).

What kinds of decision rules do women use to decide that they have been raped? A study by Linda Williams (1984) of 246 female rape victims who obtained help from a Seattle rape relief center uncovered why rape victims report their rapes to police authorities. Reporting, of course, is different from defining, but this study does give us some valuable information about how some women understand rape. The women who were the most likely to report their rapes were those who had endured what Williams called "classic rapes": They had been forcibly raped in their own homes, cars, or public places, by strangers, and they had experienced a great deal of serious injury. Women who were sexually assaulted in a social situation (e.g., at a party or on a date) by people whom they knew (acquaintances, friends, or relatives) were much less likely to report the incidents to police. Part of the reason is that a prior relationship between the rapist and his victim meant that the rape was usually less violent, caused less physical damage to the victim, and required less immediate medical aid. However, even if a woman was severely injured during a rape, she was still less likely to report the crime if she had been raped by someone whom she knew (p. 464). Williams concluded that when the elements of a classic rape are present, they provide the victim with the evidence that she needs to convince herself that she is indeed a true rape victim (Williams, 1984, p. 464).

We cannot know for sure that Williams's interpretation is the correct one. All women who are raped may know it regardless of how classic it is, but they may feel confident enough that police authorities will believe their complaints only if their rapes show the classic elements. Victims of "unclassic" rapes—forced sex by a man who has some prior relationship to them and who does not beat them or use a weapon—may see themselves as true victims but they may not see themselves as legitimate crime victims who can get justice in a court of law (Estrich, 1987). In this belief, they are far from wrong. Prior relationship rapes are more likely to be viewed by others (e.g., police and prosecutors) as private matters, the rapist is more likely to be viewed as less blameworthy, and the victim is more likely to be held partly responsible (Estrich, 1987). Even rape victims themselves report that when they do not tell authorities about their victimizations it is because they view them as private or personal matters (Bureau of Justice Statistics, 1997, pp. 94-95). Little doubt exists, however, that without the elements of a classic rape, some women will decide that they have not been raped at all. Studies of spousal rape report that wives are reluctant to use the word *rape* to describe their nonconsensual sexual experiences with their mates even when a great deal of force is present (Bergen, 1996, p. 48).

Rape and Social Conflict

Rape is inextricably tied up with the social conflicts and interpersonal hostilities that regularly characterize male/female relationships (Jackson, 1995, p. 24). These sexual antagonisms can breed contempt in which males come to hate females and vice versa. Whereas men are in a position to express their contempt for women through forcible rape, women must find other ways to vent their disdain for men (Smith & Bennett, 1985, p. 303). The lower the status of females in contrast to males, the higher the forcible rape rate (Baron & Straus, 1987, p. 481; Schwendinger & Schwendinger, 1983). Where men are both dominant and domineering, rape is one indicator of females' low status, and it is one way that women's low status is maintained and reinforced (Baron & Straus, 1987, p. 481; Baron & Straus, 1989; Funk, 1993, p. 27; Higgins & Silver, 1991, pp. 1-2). Rape does seem to be more likely in societies where women have little power in comparison to men. Reiss's (1986, pp. 191-192) statistical analysis of rape in many countries found that the belief in female inferiority tends to be positively correlated with the occurrence of rape. Social and cultural factors channel and direct sexual urges, and sometimes these factors make it more likely that some men will physically and emotionally terrorize some women (Buchwald, Fletcher, & Roth, 1993).

It is not hard to realize that cultural understandings about the causes of rape can actually precipitate certain social behaviors instead of others. The belief that an impetuous male sex drive causes men to lose control and then to rape women may itself be partly responsible for the maintenance of male dominance (Sanday, 1996, p. 26). This belief allows rapists to neutralize conventional forces of social control by disclaiming personal responsibility for what they have done (Jackson, 1995, p. 19). It also may make it easier for observers to excuse more readily a rapist's violence. Not all people in all cultures believe that men are forced to rape by some unruly sex drive, and not all people in all cultures look at rape as an impulsive act (MacKellar, 1975, p. 18). The Minangkabau peoples of West Sumatra have a fairly rape-free society. These people believe that rape is neither impulsive nor uncontrollable (Sanday, 1990, p. 192). Because humans are capable of exerting conscious control over many impelling bodily drives—hunger, thirst, defecation, urination—it is very probable that men can control their violent tendencies as well.

Defining women as property, and as sexual and seductive, can foster a sense of entitlement. Entitlement is a cultural belief that men are entitled to goods and services from women *as a class* (Bart & O'Brien, 1985, pp. 100-101). Some men rape some women because they believe that sex is something to which they are entitled whenever they want it. If it is not given freely, then they believe it is entirely proper for them to take it.

> [Sexual] violence has deep roots in sociocultural constructions of gender and heterosexuality, constructions that promote male entitlement and social and political inequality for women. Cultural norms and myths, sexual scripts and social roles link various forms of violence and deny assistance to its victims. (Koss et al., 1994, p. 17)

Sometimes a man may rape a woman, not specifically to punish or to hurt her, but because she is a representative of a class that he despises and depersonalizes (Scully, 1990). A study of the sexual violence experienced by street prostitutes traced the violence directly to cultural beliefs about them. According to prostitutes themselves, the men who raped them were acting out core beliefs: Women who are sexually experienced cannot be harmed by sexual violence; all prostitutes are alike; and prostitutes deserve whatever happens to them (Miller & Schwartz, 1995, pp. 9-16). Any woman can be become a target of sexual violence if a rapist defines her as belonging to a class whose members are promiscuous and valueless.

Rape itself may be used as a form of social control—as a way to punish a woman who is perceived to have deviated from important norms of her gender. Consider the Cheyenne Indians, one of the important tribes of the Great American Plains. The Cheyenne were famous for their chastity, and they took marriage very seriously. At puberty, females donned a chastity belt. It was a thin piece of rope that was knotted around the waist and then passed between the thighs. It was always worn at night and during the day when the woman was away from home. If a male trifled with a woman's rope, it was considered a very serious matter. He could be beaten or even killed by female or male relatives of the offended woman. His goods would be destroyed and his horses killed. Even his parents could have their property ravaged (Llewellyn & Hoebel, 1941, pp. 176-177).

If a female yielded to a man and allowed herself to be seduced, she was permanently disgraced. Her trespass would never be forgotten, and she would be scorned wherever she went. It would be almost impossible for her to marry because no man would have her (unless she was ritually purified by a shaman). If a wife was unfaithful to her husband, it was perfectly legal for him to administer what was called the "free woman penalty" or "on the prairie." The husband would invite all the unmarried men in his soldier band and all his male cousins (except for his wife's relatives) to a spot on the prairie for a "feast." He would then turn his wife over to them, and she would be raped there by 40 to 50 men. From then on, she was a woman apart who lived in permanent shame and embarrassment (Llewellyn & Hoebel, 1941, pp. 202-203). Some disagreement existed among the Cheyenne over the propriety of using a gang rape to punish an unfaithful wife, but the custom persisted and continued to be defined, at least officially, as a right of a husband against an unfaithful wife (Llewellyn & Hoebel, 1941, p. 209).

In some cultures, rape is used to punish women for things other than extra-marital sexual relationships. The Mehinaku of central Brazil strictly segregate the sexes, and the women are absolutely forbidden to enter the sacred and private communal house of the men. If women spy on the men or their sacred rituals or artifacts, they can be punished with a gang rape (Gregor, 1982). How about in the United States? Isn't rape one of the possible consequences for women of doing things that are too unconventional, untraditional, or unfeminine, such as staying out too late, hitchhiking, being out alone, dressing and acting provoca-tively, being stoned or drunk, running away from home, or holding unfeminine jobs? In the 19th-century United States, women were able to get protection from violent husbands only if they fit the Victorian image of a proper woman—some-one who was weak, dependent, and passive. Those women who refused to follow expected feminine standards were more vulnerable to sexual attack, and they had few avenues for obtaining redress if they were raped (Pleck, 1979).

Rape Scenes

The fact that violence can be initiated by the individual who ultimately ends up as its victim was first used by Wolfgang (1958) to account for some of the variation in patterns of criminal homicide. He classified a homicide as victim precipitated if an individual initiated through word or deed the fight that caused his or her own death (p. 252). Can there be victim-precipitated rape? Amir (1971) certainly thought so. He called a rape victim precipitated if a woman said or did something to cause her own rape (or *failed* to do or say something that could have stopped a rape in progress):

> The term "victim precipitation" describes those rape situations in which the victim actually, or so it was deemed, agreed to sexual relations but retracted before the actual act or did not react strongly enough when the suggestion was made by the offender(s). The term applies also to cases in risky situations marred with sexuality, especially when she uses what could be interpreted as indecency in language and gestures, or constitutes what could be taken as an invitation to sexual relations. (p. 266)

Amir concluded that 19% of the 646 rapes he studied were victim precipitated. However, any estimate of the number of victim-precipitated rapes is open to dispute, and the higher the estimate, the less credible it is (Brownmiller, 1975). We do not have consistent definitions of rape or clear decision rules for recognizing rape when we see it. How can we know how many (or *if* any) are victim precipitated?

The belief that rape can be victim precipitated is *itself* a precipitant of rape: That is, some rapes are motivated and then rationalized by the belief that they are caused by the victim (Gilmartin-Zena, 1988, p. 289). Victim blaming lets practically everybody else off the hook: actual rapists, potential rapists, and potential rape victims (Tieger, 1981). Men who rape are more likely to blame women for the sexual encounter even if it has been openly violent, and they tend to believe more strongly than other individuals that rapes are victim precipitated (Garrett-Gooding & Senter, 1987). These beliefs motivate sexual aggression toward women and then make it possible for rapists to excuse and to justify to themselves and to others what they have done (Berkowitz, Burkhart, & Bourg, 1994, p. 8).

Scully and Marolla (1984) studied the accounts of 114 incarcerated rapists in a Virginia prison who had been convicted of rape or attempted rape. Most men in the sample accounted for their rapes by discrediting their female victims and blaming them for what happened. They claimed that the women were promiscuous, seductive, or sexually aroused, and they portrayed the encounter as a consensual sex act. Even those rapists who did admit to a rape still attempted to make themselves look better by blaming what had happened on their use of alcohol or other drugs—a temporary relapse—or on a simple misunderstanding of the women's wishes. Practically everybody offered some account for what he had done. Even a rapist who had raped five victims at gunpoint and then stabbed each one to death attempted to make himself look good.

> Physically they enjoyed the sex [rape]. Once they got involved, it would be difficult to resist. I was always gentle and kind until I started to kill them. And the killing was always sudden, so they wouldn't know it was coming. (Scully & Marolla, 1984, p. 541)

The lengths to which some rapists will go to account for what they have done is truly amazing!

Vocabularies of motive, rationalizations, and accounts have a powerful impact on patterns of sexual violence and how they are viewed. Women's unsureness about how or when they can resist and men's ability to motivate and then to excuse and to justify to others what they have done are powerful determinants of the course of social relationships. An adolescent male may initially have great ambivalence about the propriety of sexually coercive behavior. In time, however, he may acquire group-based vocabularies of motive and rationalizations that are sufficient for him to overcome any victim resistance he may encounter.

> In a rape-supportive culture, attitudes and values that degrade women and make rape possible are in existence and learned at some level by all men . . . Some men choose to accept these values, integrating them into their own, and create neutralizations that allow them both to maintain their self-image as law-abiding citizens and also to victimize women. (Miller & Schwartz, 1995, p. 17)

Eventually, sexual violence may come to be viewed as reasonable and correct (Bourque, 1989, p. 289). The man devalues and depersonalizes women (some of them at any rate) and reduces them to sex objects by defining them as "loose," "teases," or "sexually experienced." Benedict (1993) claimed that English is a language of rape in that it provides a vocabulary that portrays women as sexual, seductive, and subhuman, while portraying rape as an act of passion, pleasure, or comedy instead of an act of brutality and violence (p. 103).

Each successful sexual assault that a man commits reinforces the original definition of the situation that some women can be sexually assaulted. Peer group support for sexual violence against women and acceptance of the accounts that a rapist offers make coercive sexuality even more likely (Kanin, 1967, pp. 501-503). We must remember, however, that some men who are not tightly integrated into male peer groups still come to believe that some women can be legitimately raped (Gebhard, Gagnon, Pomeroy, & Christiansen, 1965). These men may acquire these understandings from parts of culture or may simply create them on their own from personal experiences that they have had. In some cases, rape is simply a way for some men who lack sufficient levels of personal control to achieve immediate, simple gratification of their own sexual desires (Larragoite, 1994, p. 167).

Normative Sexual Violence:
Men, Women, and Relationships

At certain times, in certain places, sexual violence against women is expected, encouraged, or even trendy. Consider Ingham, a small town in the countryside of Australia. Schultz (1978) identified three characteristics of this community that made sexual violence against women a common occurrence. First, women were viewed as men's possessions. Every woman was expected to lead a virtuous life under the roof of her father until she married, whereupon she was expected to assume cheerily the duties of a faithful wife and devoted mother under the roof of her husband. Second, women were never supposed to do or to say anything that might bring shame on their families. Third, women were devalued in Ingham. They were placed into one of two mutually exclusive categories— virgin or whore—depending almost entirely on how sexually experienced they were perceived to be. The virgin received some grudging respect for her chastity, but she was still resented because she spurned the sexual advances of men. The whore was considered contemptible by both men and women of the town.

Factors worked in concert in Ingham to make it likely that women would be raped by men. Because women were viewed as men's possessions and devalued, men had little reason *not* to treat them as sex objects. This depersonalization was especially likely with regard to a woman who was defined as a whore. Because women were expected to protect their families from any kind of humiliation, any woman who was raped was under strong pressure to keep it to herself and to suffer in silence. Rape victims were unlikely to tell even their closest friends or members of their families, and they were even less likely to tell police. The women of Ingham were in an extremely difficult situation. Their upbringing encouraged them to blame themselves and to handle the shame and guilt alone if they were raped (Schultz, 1978, p. 123).

Victim selection usually took place at weekly cabarets, the center of social life in Ingham. At these parties, attended by hundreds of men and women, one member of the "rape squad" would select a victim and pump his arm up and down (as if he were pulling the cord on a train). This motion would then be copied by other men who wanted to participate. Once the selection had been made, the men waited until their intended victim left the cabaret and followed her to some convenient location; then all of them would rape her in turn. Sometimes the "rape squad" would follow a man and a woman to a secluded place. A member of the squad would tell the man that they wanted to "train his woman." If the man refused to assent, he was criticized for being unmanly and threatened with rejection in a town where manliness was highly prized and where acceptance by other males was very important. In this small town, the pressures on men to share women were almost as great as the pressures on women to be uncomplaining victims.

The Ingham case suggests that when men are strongly integrated into a cohesive, all-male group, both the devaluation of women and sexual violence against them are more likely. However, we do not have to travel all the way to

Australia to see these processes in operation. Consider the fraternity gang rape described by Sanday (1990, pp. 19-20). It occurred in February 1983, at a West Coast university, at the end of one of the weekly parties of a Greek organization identified only as Fraternity XYZ. A woman named Laurel was semiconscious in one of the bedrooms in the fraternity house. Some of the fraternity brothers viewed her as a sexual target: She was vaginally penetrated by at least five of them while others stood by and watched and cheered. It was not until 5 days later that Laurel told a university administrator that she had been raped by several men at the XYZ house. She claimed that she had not reported the rape immediately because she believed that she could handle the matter alone. Her delay in reporting, however, was taken by some people on campus to mean that she did not really believe that she had been wronged and that nothing really serious had happened.

The brothers of XYZ thought Laurel was being cruel and vindictive in an effort to get them into trouble. They defined themselves as the injured parties, and most of them thought that they had done nothing wrong. They were astounded anyone would consider what had happened a rape, and most of the brothers refused to believe anything out of the ordinary had occurred with Laurel. They branded her as the one who was really at fault because she drank too much, wanted the sex to happen, and offered no resistance to their sexual advances. She was vilified as a "nympho," "fish," and "red meat." This attack on Laurel, coupled with the brothers' legitimation of what they had done to her, renewed the bonds of brotherhood among those men who directly participated in the event, as well as among those who derived some vicarious enjoyment from it (Sanday, 1990, p. 109). Most of the members of XYZ fraternity continued to believe that what had happened was all in fun, that Laurel had brought it upon herself, and that all the participants were too drunk to know any better (Sanday, 1990, pp. 68-69). Members of the fraternity who voiced an objection to the victimization of Laurel ran the risk of being labeled as "wimps," "gays," or "faggots."

How was this incident viewed outside the fraternity? Did anyone believe a rape had occurred? Laurel certainly did. She sued the state, the university, the fraternity, and the sorority she had been pledging at the time for $2.5 million. A year after the rape (in June 1987), the fraternity's charter was revoked by its national organization. Feminists on campus who learned of the incident also believed that a rape had occurred. Their definition of the situation was supported by the local District Attorney of the Sex Crimes Unit. Sanday, an expert on the case, also believed a rape had occurred.

If the incident had been officially branded as a rape, it would have had serious repercussions. The men responsible could have been prosecuted and imprisoned for many years, and the university would have had to have taken strong measures against the fraternity and the entire Greek system. In the end, the university administration took the path of least disruption. The offending members of the fraternity were required to take a reading course to increase their understanding of feminist issues and to perform several hours of community service. The fraternity house itself was closed for one semester (Sanday, 1990, p. 62).

Though not all men in all fraternities are rapists, fraternities are organizations that do make the sexual coercion of women a strong possibility. Kanin concluded back in 1967 from his study of coercive sexuality on a university campus that college fraternities attracted sexually aggressive males and then supported or even championed a conquest mentality toward women. The fraternities emphasized the goal—an accumulation by members of acts of sexual intercourse with females—with much less attention directed to the means to achieve the goal. When persuasion failed, a male felt comfortable in using force to get what he wanted (Kanin, 1967). Fraternities create a climate in which the use of coercion to reach sexual goals is normative and few restraints exist to curb this pattern of behavior (Yancey & Hummer, 1989).

Acquaintance rapes are the bulk of all rapes, and date rape is a type of acquaintance rape that grows out of a social or romantic situation (Hall, 1995, p. 83). In some instances, sexual assaults on campuses occur because of indifference or negligence on the part of college and university administrators (Bohmer & Parrot, 1993, pp. 10-11). They find it easier to stick their heads in the sand than to deal directly with causes, offenders, and victims. In other instances, rape on campus—and disputes over what it is, how many rapes exist, and how they should be handled—reflects some of the natural ambiguity in dating relationships (Ward, Dziuba-Leatherman, Stapleton, & Yodanis, 1994). "Sexuality is messy, passionate, unclear, tentative, anxiety-producing, liberating, frightening, embarrassing, consoling, appetitive, and cerebral. In other words, sexuality is contradictory, it is different for different people, and it is even different for the same person at different times" (Schwartz, 1995, p. 35). The ambiguity of human sexuality is one of the principal justifications for the formulation of what came to be called the Antioch Policy (at Antioch College in Yellow Springs, Ohio). This policy requires that each new level of intimacy be verbally consented to by a partner. Without this consent, any subsequent sex is viewable as forcible and can be penalized through one of several sanctions: expulsion, suspension, mandatory therapy, or restrictions on class schedulings

(Antioch College Community, 1995). More than ever before, males are expected to take steps to be certain that they are not forcing themselves on their sexual partners and that the sexual encounters in which they are involved are mutually rewarding (Pineau, 1996, pp. 17-18).

Does a rape crisis currently exist at colleges and universities? It all depends on what is meant by *rape*. Some researchers use such a broad definition of sexual violence (and accept any claim of sexual victimization without question) that the difference between the violence of rape and the misfortune of bad sex is lost in the shuffle. Roiphe (1995) insisted that the claim that a rape epidemic now exists on college and university campuses is caused more by a new way of interpreting or seeing than it is by actual changes in some physical reality.

> We all agree that rape is a terrible thing, but we no longer agree on what rape is. Today's definition has stretched beyond bruises and knives, threats of death or violence to include emotional pressure and the influence of alcohol. (p. 75)

The desire by women to free themselves from male-generated sexual pressures in order to increase their freedom and independence is reasonable. However, broadening the definition of rape to include any bad sex—Cathy Young (1992) branded this as "definitional shenanigans"—may actually lead to the unintended outcome of trivializing severe forms of sexual violence.

Research has found that most males who commit date rape do not define it as rape (Capraro, 1994, p. 22). Male college students and males in general tend to believe that a woman's refusal to engage in sex is usually nothing more than a required part of the dating ritual (Celes, 1991). Males rarely view a "no" from their dates as being as definite as they do a "no" from a boss or a loan officer. A "no" is even more likely to look meaningless to an individual who truly believes that sex should be the outcome of any male/female relationship and that it is one of the great pleasures of life.

It is possible to have a great deal of sexual violence in a relationship without the encounter automatically being defined as forcible rape. Robert LeVine (1959) studied sexual relationships between the men and women of Gusii, a small agricultural community in the highlands of southwestern Kenya in Africa (pp. 965-990). Every relationship between males and females in this community was filled with hostility, and even marriage brought no relief from sexual antagonisms. All it did was to bring together a man and a woman from different and probably warring clans and unite them in a continually stormy relationship. Women were especially ambivalent about marriage. They needed to marry because marriage and the birthing of sons were the only ways that women could gain a measure of status and respect. However, marriage forced them to sever ties with their own families and to live with a clan that might be cruel and mean-spirited. Practically every Gusii bride was fearful of what would happen to her when she left the protection of her own home and family.

Even though a woman's parents had granted permission for the marriage to proceed (and the woman herself had adjusted to the prospect), the bride-to-be strenuously resisted when her husband-to-be sent escorts to get her. Part of the resistance was obligatory; she could not look too willing, or she might be considered a trollop. However, much of the reluctance was genuine and born out of anxiety over just what would become of her in her future husband's clan. She might run away and hide. Her parents might have to find her and convince her to go. She might have to be dragged out of her house by her cortege, all the while kicking and screaming. She went to her new home in tears, with her hands on top of her head, in a show that combined sorrow, regret, and a fear of the unknown (p. 968).

The wedding night sexual encounter was an ordeal for both husband and wife, and it displayed the kinds of sexual conflicts that plagued all Gusii marriages. The bride put her new husband's sexual abilities to the test. She did everything that she could to ensure his impotence and to guarantee his failure at intercourse. She might use magical charms. These included things such as chewing a small piece of charcoal, placing a piece of knotted grass under the marriage bed, or twisting the phallic flower of the banana tree. She might knot her pubic hairs so that they covered the opening of her vagina, or she might constrict her vaginal muscles so tightly that any penetration by her husband would be impossible. She might even refuse to get onto the bed. Her resistance to his advances persisted throughout the entire wedding night, and wives took great enjoyment in being able to keep their husbands from achieving intercourse for an indefinite period of time.

The new husband did all that he could to guarantee his own success and to overcome his wife's resistance. He made certain that he was well fed by eating special herbs as well as large quantities of coffee beans, valued as an aphrodisiac. He was cheered on by his brothers and male cousins, who stayed close to the wedding-night dwelling and sang and danced while the events unfolded. The man's relatives might become more than passive onlookers. If the new groom was young (under 25) and his new bride was particularly uncooperative, some of these men might come to the rescue. They might grab her, undress her, throw her on the bed, and hold her down until the husband could penetrate her. The husband wanted to cause his new bride as much physical pain as possible. In fact, his status was greatly enhanced if she was unable to walk the next day. "Legitimate heterosexual encounters among the Gusii are aggressive contests, involving force and pain-inflicting behavior which under circumstances that are not legitimate could be termed 'rape' " (p. 971). Even though husbands obtained sex from their wives through force and wives strenuously resisted, the sex was not defined as rape. This was true even though the entire encounter was intentionally painful for the woman.

Summary and Conclusions

This chapter has examined sexual violence, rape, and sexual assault. Rape is primarily a way for certain individuals to enhance and to reaffirm their sense of power, potency, and control by hurting and disempowering others. We have seen that what qualifies as rape in one place and time may be treated very differently in another place and time, regardless of how damaging or injurious the assault may be to its victim. Some of the factors used to define rapes are the amount of force that is present during some sexual relationship and how much resistance a victim offers. We have seen that patterns of sexual violence and attitudes about it are powerfully influenced by social factors—culture and subculture, gender-based associations, sexual inequality, ideologies, symbols, motives, and accounts. Structured inequality between males and females may lead to conflicts. Sex may come to be viewed by some men as something that they have a right to receive from women, so that if it is not freely given, they have no qualms about using force to get it. They may then attempt to excuse and to justify what they have done to others or even to themselves. Times and places may be found where violence against women is normative and customary.

References

Allison, J., & Wrightsman, L. (1993). *Rape, the misunderstood crime.* Newbury Park, CA: Sage.
Amir, M. (1971). *Patterns in forcible rape.* Chicago: University of Chicago Press.
Antioch College Community. (1995). Antioch College: A sexual consent policy. In B. Leone & K. Koster (Eds.), *Rape on campus* (pp. 10-22). San Diego, CA: Greenhaven.
Barlow, H. (1996). *Introduction to criminology* (7th ed.). New York: HarperCollins.
Baron, L., & Straus, M. (1987). Four theories of rape: A macrosociological analysis. *Social Problems, 34,* 467-489.
Baron, L., & Straus, M. (1989). *Four theories of rape in American society: A state-level analysis.* New Haven, CT: Yale University Press.
Bart, P., & O'Brien, P. H. (1985). *Stopping rape: Successful survival strategies.* New York: Pergamon.

Bedau, H. A. (1982). The death penalty for rape: *Coker v. Georgia* (1977). In H. A. Bedau (Ed.), *The death penalty in America* (3rd ed., pp. 299-304). New York: Oxford University Press.

Benedict, H. (1993). The language of rape. In E. Buchwald, P. Fletcher, & M. Roth (Eds.), *Transforming a rape culture* (pp. 101-105). Minneapolis, MN: Milkweed.

Bergen, R. K. (1996). *Wife rape: Understanding the response of survivors and service providers.* Thousand Oaks, CA: Sage.

Berkowitz, A., Burkhart, B., & Bourg, S. (1994). Research on college men and rape. In A. Berkowitz (Ed.), *Men and rape: Theory, research, and prevention programs in higher education* (pp. 3-19). San Francisco: Jossey-Bass.

Bohmer, C., & Parrot, A. (1993). *Sexual assault on campus: The problem and the solution.* New York: Lexington.

Bourque, L. B. (1989). *Defining rape.* Durham, NC: Duke University Press.

Brownmiller, S. (1975). *Against our will: Men, women, and rape.* New York: Simon & Schuster.

Buchwald, E., Fletcher, P., & Roth, M. (1993). *Transforming a rape culture.* Minneapolis, MN: Milkweed.

Bureau of Justice Statistics. (1997). *Criminal victimization in the United States, 1994.* Washington, DC: Government Printing Office.

Capraro, R. (1994). Disconnected lives: Men, masculinity, and rape prevention. In A. Berkowitz (Ed.), *Men and rape* (pp. 21-33). San Francisco: Jossey-Bass.

Carosella, C. (1995). Introduction. In C. Carosella (Ed.), *Who's afraid of the dark? A forum of truth, support, and assurance for those affected by rape* (pp. xiii-xxiv). New York: Harper Perennial.

Celes, W. (1991, January 2). Students trying to draw line between sex and assault. *New York Times,* p. A1.

Chappell, D. (1976). Cross-cultural research on forcible rape. *International Journal of Criminology and Penology, 4,* 295-304.

Coker v. Georgia, 433 U.S. 485 (1977).

Cuklanz, L. (1996). *Rape on trial: How the mass media construct legal reform and social change.* Philadelphia: University of Pennsylvania Press.

Epstein, J., & Langenbahn, S. (1994). *The criminal justice and community response to rape.* Washington, DC: Government Printing Office.

Estrich, S. (1987). *Real rape.* Cambridge, MA: Harvard University Press.

Federal Bureau of Investigation. (1998). *Crime in the United States 1997.* Washington, DC: Government Printing Office.

Ferraro, K. (1996). Women's fear of victimization: Shadow of sexual assault? *Social Forces, 75,* 667-690.

Fitzgerald, L. (1993). Sexual harassment: Violence against women in the workplace. *American Psychologist, 48,* 1070-1076.

Fried, S. (1997). The rapist. In L. Salinger (Ed.), *Deviant behavior 97/98* (2nd ed., pp. 39-43). Guilford, CT: Dushkin.

Funk, R. (1993). *Stopping rape: A challenge for men.* Philadelphia: New Society.

Garrett-Gooding, J., & Senter, R., Jr. (1987). Attitudes and acts of sexual aggression on a university campus. *Sociological Inquiry, 57,* 348-371.

Gebhard, P., Gagnon, J., Pomeroy, P., & Christiansen, C. (1965). *Sex offenders: An analysis of types.* New York: Harper & Row.

Gilmartin, P. (1994). *Rape, incest, and child sexual abuse.* New York: Garland.

Gilmartin-Zena, P. (1988). Gender differences in students' attitudes toward rape. *Sociological Focus, 21,* 279-292.

Goode, E. (1996). Rape. In E. Goode (Ed.), *Social deviance* (pp. 284-287). Boston: Allyn & Bacon.

Goodman, L., Koss, M., Fitzgerald, L., Russo, N., & Keita, G. (1993). Male violence against women: Current research and future directions. *American Psychologist, 48,* 1054-1058.

Gordon, M., & Riger, S. (1989). *The female fear.* New York: Free Press.

Greenfeld, L. (1997). *Sex offenses and offenders.* Washington, DC: Government Printing Office.

Gregersen, E. (1983). *Sexual practices: The story of human sexuality.* New York: Watts.

Gregor, T. (1982). No girls allowed. *Science Magazine, 3,* 26-31.

Groth, N. (1979). *Men who rape.* New York: Plenum.

Hall, R. (1995). *Rape in America.* Santa Barbara, CA: ABC-CLIO.

Higgins, L., & Silver, B. (1991). Introduction: Rereading rape. In L. Higgins & B. Silver (Eds.), *Rape and representation* (pp. 1-11). New York: Columbia University Press.

Jackson, S. (1995). The social context of rape: Sexual scripts and motivation. In P. Searles & R. Berger (Eds.), *Rape and society* (pp. 16-27). Boulder, CO: Westview.

Kanin, E. J. (1967). Reference groups and sex conduct norm violations. *Sociology Quarterly, 8,* 495-504.

Koss, M. (1993). Rape: Scope, impact, interventions, and public policy responses. *American Psychologist, 48,* 1062-1069.

Koss, M., Goodman, L., Browne, A., Fitzgerald, L., Keita, G. P., & Russo, N. P. (1994). *No safe haven: Male violence against women at home, at work, and in the community.* Washington, DC: American Psychological Association.

Larragoite, V. (1994). Rape. In T. Hirschi & M. Gottfredson (Eds.), *The generality of deviance* (pp. 159-172). New Brunswick, NJ: Transaction.

LeVine, R. (1959). Gusii sex offenses: A study in social control. *American Anthropologist, 61,* 965-990.

Llewellyn, K. N., & Hoebel, E. A. (1941). *The Cheyenne way: Conflict and case law in primitive jurisprudence.* Norman: University of Oklahoma Press.

Macdonald, J. (1995). *Rape.* Springfield, IL: Charles C Thomas.

MacKellar, J. (1975). *Rape: The bait and the trap.* New York: Crown.

McKinney, K. (1994). Sexual harassment and college faculty members. *Deviant Behavior, 15,* 171-191.

Miller, J., & Schwartz, M. (1995). Rape myths and violence against street prostitutes. *Deviant Behavior, 16,* 1-23.

Palmer, C. (1989). Is rape a cultural universal? A re-examination of the ethnographic data. *Ethnology, 28,* 1-16.

Pelka, F. (1995). Raped: A male survivor breaks his silence. In P. Searles & R. Berger (Eds.), *Rape and society* (pp. 250-256). Boulder, CO: Westview.

Pineau, L. (1996). Date rape: A feminist analysis. In L. Francis (Ed.), *Date rape* (pp. 1-26). University Park: Pennsylvania State University Press.

Pleck, E. (1979). Wife beating in nineteenth-century America. *Victimology: An International Journal, 4,* 60-74.

Reiss, I. L. (1986). *Journey into sexuality.* Englewood Cliffs, NJ: Prentice Hall.

Roiphe, K. (1995). A critique of "rape-crisis" feminists. In B. Leone & K. Koster (Eds.), *Rape on campus* (pp. 74-83). San Diego, CA: Greenhaven.

Sanday, P. R. (1990). *Fraternity gang rape: Sex, brotherhood, and privilege on campus.* New York: New York University Press.

Sanday, P. R. (1996). *A woman scorned: Acquaintance rape on trial.* New York: Doubleday.

Schultz, J. (1978). Appendix 1: The Ingham case. In P. Wilson (Ed.), *The other side of rape* (pp. 112-125). Queensland: University of Queensland Press.

Schwartz, P. (1995). A negative view of the Antioch Plan. In B. Leone & K. Koster (Eds.), *Rape on campus* (pp. 33-38). San Diego, CA: Greenhaven.

Schwendinger, H., & Schwendinger, J. (1983). *Rape and inequality.* Beverly Hills, CA: Sage.

Scully, D. (1990). *Understanding sexual violence: A study of convicted rapists.* Boston: Unwin Hyman.

Scully, D., & Marolla, J. (1984). Convicted rapists' vocabulary of motive: Excuses and justifications. *Social Problems, 31,* 530-544.

Smith, M. D., & Bennett, N. (1985). Poverty, inequality, and theories of forcible rape. *Crime and Delinquency, 31,* 295-305.

Spohn, C., & Horney, J. (1992). *Rape law reform: A grassroots revolution and its impact.* New York: Plenum.

Svalastoga, K. (1962). Rape and social structure. *Pacific Sociological Review, 5,* 48-53.

Tieger, T. (1981). Self-rated likelihood of raping and the social perception of rape. *Journal of Research in Personality, 15,* 147-158.

Ward, S., Dziuba-Leatherman, J., Stapleton, J. G., & Yodanis, C. (1994). *Acquaintance and date rape: An annotated bibliography.* Westport, CT: Greenwood.

Weis, K., & Borges, S. (1973). Victimology and rape: The case of the legitimate victim. *Issues in Criminology, 8,* 71-115.

Williams, L. (1984). The classic rape: When do victims report? *Social Problems, 31,* 459-467.

Wolfgang, M. (1958). *Patterns in criminal homicide.* Philadelphia: University of Pennsylvania Press.

Yancey, P., & Hummer, R. (1989). Fraternities and rape on campus. *Gender and Society, 3,* 457-473.

Young, C. (1992, May 31). Women, sex and rape: Have some feminists exaggerated the problem? *Washington Post,* p. C1.

6

Suicide

Introduction: Organizing the Death Experience

Suicide is a term usually used to refer to an intentional killing of oneself by oneself for the sole purpose of ending one's life and existence. The term is, in all likelihood, based on the Latin pronoun *sui,* "self," and the Latin verb *cide,* "to kill" (Farberow, 1975, p. 1). According to the *Oxford Dictionary,* the term was first used in English in 1651, but it actually appeared a bit earlier in a work by Sir Thomas Browne, published in 1642 (Alvarez, 1972, pp. 50-51). Before the mid-1600s, self-killings were treated as murders rather than placed in a category all their own.

Why did the term *suicide* appear when it did? Throughout the Middle Ages (476 A.D. to 1450 A.D.), practically everyone believed, fervently and without question, in the influence and power of supernatural forces and spiritual beings. Heaven and hell were real places, and their existence exerted a shadow over practically everything that everybody did. With this worldview, it was impossible for anyone—certainly for anyone with even a hint of religious belief—to commit a self-inflicted killing to end all existence and send himself or herself into nothingness. A self-inflicted death, it was believed, would simply move one to a different realm of existence—perhaps better, perhaps worse—but an existence nonetheless. The secularization of human experience that occurred in the years following the Middle Ages promoted a worldview that had less room for spiritual matters. Perhaps heaven and hell did not exist, maybe humans were physical beings without souls, and just possibly death might mean the end of all existence. For the first time, suicide became a real possibility (Shneidman, 1985, pp. 9-10).

No universal agreement exists on the significance or importance of a suicide. In some places, at some times, for some groups, any individual's death is seen as a tragedy—a diminishing of everybody else, regardless of how it occurs. It is a time of sadness; survivors themselves feel responsible for the death; and despair prevails. In other places, at other times, a suicide is viewed as a rather trivial event because death itself is unimportant. What matters is how one lives—how completely and how productively—and not how one dies (Fedden, 1938/1980, p. 13).

Suicides may seem to be private acts of desperation, but they are actually highly social, being just one of many possible outcomes of social relationships. The essence of suicide is its relativity and variability. Suicide is characterized by a diversity of motives, methods, forms, rationalizations, settings, and circumstances; the only common factor seems to be that these deaths have all been called or considered suicides by somebody. We must remember that the classification of a death as a suicide is a social process—as is the process of dying itself—and must be understood in its own right. Suicide is not an automatic classification that logically flows from a death scene.

Social Frames and the Definition of the Situation

People can die in many ways: naturally, from accidents, from murders, and from suicides. The determination that a death is a suicide hinges on judgments about the inner experiences of a dead individual. Because that person is unable to tell us anything directly about his or her motive and intent, speculation reigns. Even when a suicide note is present—and it usually is not—it is not irrefutable proof that a suicide has occurred. After all, a clever murderer could certainly leave a suicide note if he or she wanted to muddy the waters and make it more difficult for authorities to know what really happened.

In 1974, Erving Goffman explored the importance of frame analysis for understanding human behavior. For him, frames were definitions of situations that individuals in a social situation develop in order to give some meaning and organization to their collective experiences. Framing is a process of constructing images that help people to interpret events and then passing these images on to others. Framing always involves selection, presentation, and accentuation, and frames always package social experiences for *both* producers and users of them

(Papke, 1987, p. xvii). Frames help people who are confronting some hard-to-understand or ambiguous event to make some sense of it (Goffman, 1974, p. 30).

Framing a death as a suicide is hardly a straightforward enterprise. A self-inflicted, intentional death seems to be a rejection of the entire group and a refusal to find value in life by an individual who may seem to have everything that makes life worth living. A great deal of reluctance may be found among the living to frame a death as a suicide if any other option exists. Alvarez (1972, p. 87) offered an instructive example of how this can work. A West of Ireland coroner decided that a man who had shot himself in the head with his gun had actually died accidentally. How could this seemingly clear example of a suicide be framed as an accident? The coroner concluded that the gun had discharged inadvertently as the man cleaned the barrel with his tongue!

Death is not the only possible goal of a suicidal act, and the presence of a self-inflicted death is not a perfect indicator of suicidal intent. Some people who do not wish to die still do very dangerous and risky things, and not all suicidal deaths can reasonably be viewed as intended (Canetto & Lester, 1995, p. 4). Hendin (1995b, p. 237) informed us that the spark that generates a suicide is usually transient and that a large number of people who commit suicide are actually ambivalent about it. They do something in an agitated and perturbed state that commits them to a plan of action that is irreversible and deadly.

The Evolutionary Setting of Suicide

Insurmountable obstacles are present in deciding if suicides are found among other members of the animal kingdom. We do not have enough examples of self-destructive acts in natural settings to speak about them with any confidence, and those examples that we do have are of an indeterminate status—that is, we do not know that the animal intentionally killed itself, knowing the consequences. Lemmings march into the waters off the Norwegian coast in what seems to be a clear example of a group suicide, and thousands drown. However, the deaths may be due to an error in judgment rather than to some innate death wish. If they come to a modest body of water, they seem to be able to cross it with little trouble. However, if the body of water is too large to cross, then they drown. What looks like group suicide probably is not.

Knowing the inner state of an organism that commits a self-destructive act is difficult, regardless of the particular species to which it belongs. Consider the case of the male Australian redback spider (Andrade, 1996, pp. 70-71). After he sticks his sex organ into the female, he wiggles his abdomen in front of her and then positions it directly over her mouthparts. His mate then eats him alive. Why would he go out of his way to be devoured by his sexual partner? It seems that this "male copulatory suicide" is actually adaptive from the standpoint of reproductive success. While she is busy consuming him, he is busy impregnating her successfully. His act of self-sacrifice maximizes the chances that his genes, and his alone, will be passed on to future generations.

Clearly, life-threatening behaviors do occur among nonhumans, some of them leading directly to death (deCatanzaro, 1981, p. 44). However, most of these deaths either are incidental to other necessary activities such as eating, reproduction, or protection, or actually contribute to the overall fitness of the group or community—a percentage die (usually the weakest, oldest, or sickest) so that the rest may live better lives and develop healthier offspring. When a death of a nonhuman is a direct response to stress or isolation, it almost always is a response to living in a captive environment around humans, and the death is a poor basis for deciding if nonhuman animals commit suicide.

At this point in time, the data are still too sparse for us to be able to decide beyond a shadow of a doubt that nonhumans kill themselves in acts of suicide. What is clear is that most, if not all, of the factors responsible for changing suicide rates—increases in depression, increases in the use of chemical substances by the young, lowering of the age of puberty, an abundance of social stressors, greater tolerance for suicidal behaviors, increases in suicidal role models (Diekstra & Garnefski, 1995)—are really found only at the human level. Durkheim's view may still be the correct one. He believed that only humans have the intelligence to formulate an anticipatory understanding of their own death and that only humans encounter the kinds of social conditions that cause suicide (Durkheim, 1897/1951, pp. 44-45).

The Meaning of Suicide

A definition of suicide—one that has stood the test of time—was offered over 100 years ago by Emile Durkheim (1897/1951) in his book *Suicide*: "Suicide is applied to all cases of death resulting directly or indirectly from a positive or negative act of the victim himself [herself], which he [she] knows will produce this result" (p. 44). Durkheim was confident that his definition of suicide would allow all dispassionate observers to recognize suicides when they saw them and to differentiate them consistently from other forms of death (p. 42). He rejected the elements of motive and intent (p. 43) because he believed that they were not objective enough to be known by others. Even social actors themselves may misunderstand the true reasons for their own acts (p. 43). The defining event of suicide for Durkheim was the renunciation of existence and the purposeful sacrifice of life: "When resolution entails certain sacrifice of life, scientifically this is suicide" (p. 44).

Durkheim's aim to construct a purely objective definition of suicide was laudable but unrealistic (Pope, 1976, pp. 10-11). It is impossible clearly and uniformly to separate every suicidal death from all others. Suicidal phenomena do not exist independently of human thought about suicide, and a great deal of subjectivity enters into the classification of a death as a suicide (Taylor, 1990). The difference in suicide rates of Catholics and Protestants in the Netherlands during the years 1905 to 1910 was directly related to a classification bias. A large proportion of the deaths of Catholics that had been recorded as "death from ill-defined or unspecified cause" or "sudden death" would have been recorded as suicides if the deceased had been Protestants (Poppel & Day, 1996, pp. 505-506).

Durkheim confused motivation (the reasons people kill themselves) with intent (a knowledge or awareness of possible outcomes or consequences), and motive and intent worked their way into his analysis of suicide. He apparently never considered the implications for his analysis of using instances of suicide that were determined by coroners and medical examiners, police, judges, and other authorities, who undoubtedly based their decisions on all those subjective variables that he had discarded (Pope, 1976, p. 11). The "intrinsic ambiguity" of suicide recognized by Douglas (1967, p. 251) exists because human actions and reactions are contextual, and the meaning of any social event requires observers to infer and to speculate about a great many details.

Suicide has distinctive meanings for an individual contemplating it (Douglas, 1967, pp. 284-319). It gives an individual the prospect of starting over again, albeit in a different sphere of existence, as a new being. It produces changes in how one is viewed by others, and it definitely changes human relationships (and produces an immediate *re*interpretation of existing relationships by surviving acquaintances, friends, and family members). Suicide also offers an opportunity to avenge some actual or imagined injustice. These social meanings exert a powerful influence over the course and the interpretation of self-destructive acts (Douglas, 1967, p. 321). On the island of Tikopia, in the western Pacific, individuals take suicide swims out into the ocean. These swims are one legitimate way for them to handle life's problems. The swimmers have time alone in a watery world to think, and these swims actually enhance their standing in the community. Community ties are reaffirmed because the swimmers allow themselves to be rescued and returned to shore (Littlewood & Lipsedge, 1987, p. 292).

A suicidal person must formulate appropriate motives for the self-destructive act, work to accomplish his or her goal of dying, and neutralize any constraints that he or she may have against suicide. People who successfully kill themselves may come to believe that they have an intolerable problem in living, not of their own making, which they cannot solve. Suicide comes to be viewed as a reasonable, perhaps the only, solution to their problem. They convince themselves that suicide will make all the troubles go away and that a better existence will be found in the hereafter. After convincing themselves that no reason exists to feel guilty or embarrassed about a self-inflicted death, they are prepared to die (Jacobs, 1970). Suicide is more likely for people who find themselves in an ambiguous situation in which suicide seems to be the only meaningful choice (Rodgers, 1995, p. 120).

People can change their views of suicide in light of events such as the suicides of others or perhaps even their own suicide attempts. Consider Alvarez's own efforts to self-destruct. He took 45 barbiturates and waited for death, but his wife discovered him and called an ambulance. He was in a deep coma, and he had a rapid pulse, vomit in his mouth, and blue skin (cyanosed). Medical technicians pumped his stomach and got him breathing again. He was unconscious for 3 days but finally recovered. He had the following to say about his experience:

> I thought death would be like that: a synoptic vision of life, crisis by crisis, all suddenly explained, justified, redeemed, a Last Judgment in the coils and circuits of the brain. Instead, all I got was a hole in the head, a round zero, nothing. I'd been swindled. (Alvarez, 1972, p. 282)

As a result of his experience with suicide, he concluded that when his death finally came, it would be nastier than suicide and a lot less convenient (Alvarez, 1972, p. 284).

Suicidal deaths present a social situation offering extreme opportunities for social discomfort. Those individuals who must cope with the suicidal death of a family member or friend may find that outsiders make the healing process more difficult. Some people hold negative views of suicide, and they may blame relatives or even friends of the decedent for what happened. The general social context is far less supportive when the death is a suicide, and even those people who wish to offer aid and comfort to surviving family members may find themselves paralyzed by a lack of clarity over just what should be done or said (Calhoun & Selby, 1990, pp. 222-223).

Ambiguity and Suicide: The Curious
Case of Autoerotic Asphyxiation

The equivocal nature of suicide is especially clear in the case of deaths resulting from a rather curious form of human activity known as autoerotic asphyxiation. Autoerotic fatalities (also called erotic asphyxiations) are deaths that occur in the context of solitary sexual activity from oxygen deprivation. (The technical name for restriction of air flow for the purpose of sexual arousal is *hypoxyphilia.*) Typically, the deceased is found alone, wearing physical restraints such as handcuffs, gags, or blindfolds, behind a locked door, with a rope or other ligature around his or her neck. Oxygen deprivation may also have been achieved through the use of plastic bags, chemicals, chest compression, or a tub or water in which the face was immersed. Evidence of masturbatory activities exists, and one goal in autoerotic asphyxiation is to heighten the pleasure of orgasm (Evans & Farberow, 1988, p. 26). When found, the decedent is completely or partially naked, and erotic materials may be present at the scene. Sexual asphyxia is most often practiced by young, white, middle- or upper-middle-class males (Lowery & Wetli, 1982). Data collected from England, the United States, Canada, and Australia show that one to two erotic asphyxiations per million population are detected and reported each year (American Psychiatric Association, 1994, p. 529).

The reason that some people engage in erotic asphyxiation is that it is supposed to be sexually stimulating. The carotid arteries (on either side of the neck) carry oxygen-rich blood from the heart to the brain. When these are compressed, as in strangulation or hanging, the sudden loss of oxygen to the brain and the accumulation of carbon dioxide can increase feelings of giddiness, lightheadedness, and pleasure, all of which will heighten masturbatory sensations. Autoerotic asphyxia was described in writing as far back as the 1600s, where it was proclaimed to be a cure for male impotence. The basis of this belief may well have been the observation that men who were executed by hanging attained an erection (Uva, 1995, p. 575). Men in Victorian England could visit "Hanged Men's Clubs," where prostitutes would administer enough asphyxiation for these men to be sexually satisfied, but not so much that they would be in any danger (Uva, 1995, p. 575).

It is hard to believe that erotic asphyxiates are unaware of the riskiness of their chosen sexual practice. The carotid arteries are quite sensitive, and even a minor miscalculation can lead to unconsciousness and death. A pull of as little as 7 pounds will diminish the flow of blood through the carotid artery sufficiently that unconsciousness results within 7 seconds. Even if death does not result, brain damage still can occur if the oxygen supply is cut off for too long (Resnik, 1972,

p. 11). Though the precise details of what can happen may be unknown to the practitioners of erotic asphyxia, they still must know something about the great risk involved. If the line between suicide and everything else hinges exclusively on the distinction between "intends to die" and "knows death could result," it is a very fine line indeed. The dangers seem to be understood by adult practitioners. Autoerotic asphyxia is called "terminal sex" in the adult bondage subculture (Uva, 1995, p. 578), but it could just as easily be called sexual termination.

By its very nature, suicide defies classification no matter what the circumstances. Decision makers may misinterpret clues at the death scene. A suicide note, if present, is usually taken as prima facie evidence that the death is a suicide. However, the suicide note itself may be part of an autoerotic game—a prop to aid the decedent's fantasy that it is possible to "die" and then be "reborn" (Hazelwood, Dietz, & Burgess, 1982, pp. 766-767). Another possibility is that a homicide will be disguised as an accident or a suicide. In fact, Wright and Davis (1976) reported just such an occurrence. A 31-year-old, single male was found hanging by his neck from a necktie tied to a shower curtain rod in his apartment. His feet touched the floor. He was blindfolded with another necktie, a belt was around his neck, and his hands had been tied behind his back. He was fully clothed except that he wore no shoes. Some of his personal possessions were missing, and two females had been seen leaving his apartment. The death scene and subsequent autopsy provided enough evidence to suggest that the death was an autoerotic asphyxiation. However, it was finally decided that the two females had actually robbed and murdered the man. An important clue was the fact that the victim's hands were tied behind his back, a seeming impossibility if the autoerotic activities were completely solitary. However, even here we must be cautious. Because some people do practice sexual asphyxia with others around, the man's hands could have been tied by a sexual partner.

Sociocultural Variations in Suicide: Motive, Intent, and Interpretation

Historical Meanings

Purposeful self-killings are not unusual, and they have been found at most times and in most places. Some of these self-inflicted lethal acts reflect an individual's isolation from social groups and are a response to loneliness, hopelessness, helplessness, depression, and despair. Other suicides are caused by an overidentification with significant groups (Durkheim, 1897/1951). Henry and Short (1954) argued that increases in the levels of frustration in a society have an important influence on suicide. If an individual who is isolated and separate from others is exposed to deeply frustrating experiences, suicide becomes more likely because the only available target of any aggression is the individual himself or herself. If an individual takes personal responsibility for his or her own frustrations and concludes that these frustrations are uncontrollable and likely to continue indefinitely, feelings of helplessness and hopelessness are more likely. This constellation of factors will make suicide more likely (Unnithan, Huff-Corzine, Corzine, & Whitt, 1994, pp. 96-98).

If suicides are a constant feature of human existence, the content of the social reactions to them is not. Evaluations of suicide range (or have ranged) from strong opposition, through tolerance and acceptance, to spirited support if the suicide was done in the proper way and for all the right reasons. In some places, the living were promised harsh penalties if they took their own lives—the corpse of anyone who committed suicide was abused, and the family name of the decedent was dishonored. At other places, the decedent was highly admired for committing suicide, and it was a death that he or she freely chose because of the benefits that it offered. The many forms of suicide, the many reasons it occurs (in terms of motive and intent), the many methods people use to self-destruct, the wide array of the social meanings of suicide, and the multifaceted nature of social reactions to suicide all make for a great deal of flexibility and variability in regard to the process of self-destruction.

Death in ancient Egypt was considered to be a passage from one form of existence to another. The living and the dead were viewed as not too different and were credited with having some of the same needs and interests. Because death was defined as no big deal, it really did not matter to anyone exactly what caused it. However, if an Egyptian who committed suicide were magically transported to another culture just as the last breath left his body, attitudes would have been different:

> If [an] ancient Egyptian had killed himself in pre-Christian Scandinavia, for example, he would have been guaranteed a place in Viking paradise. If he had taken his life during the Roman Empire, his death would have been honored as a glorious demonstration of his wisdom. If he had cut open his stomach in feudal Japan, he would have been praised as a man of principle. If he had killed himself in fifteenth-century Metz, however, his corpse would have been crammed into a barrel and floated down the Moselle, sending the tainted body beyond city limits. In seventeenth-century France his corpse would have been dragged through the streets, hanged upside down, then thrown on the public garbage heap. In seventeenth-century England his estate would have been forfeited to the crown and his body buried at a cross-roads with a stake through the heart. (Colt, 1991, p. 131)

Some groups fervently believe that suicide is a good and noble death, whereas other groups, with equal vehemence, view it as a sign of biological dysfunction, immorality, or psychological abnormality. These meanings are not without consequence for the cause, course, and effects of self-inflicted deaths and how deaths are categorized and interpreted.

The Vikings of pre-Christian Scandinavia viewed suicide as a good way to die because it was a ticket into Valhalla, the eternal home of all heroes who had died violently. The surest way to get into Viking heaven was to die heroically in battle, but a self-inflicted violent death came in a close second. The Vikings believed that people who entered Valhalla would fight mock battles and then feast for all time, drinking from the skulls of their dead enemies. The founder of the feast was none other than Odin, the Scandinavian god of war, who was supposed to have committed suicide himself. Self-respecting Vikings who saw no prospect of dying in battle often killed themselves with their swords or jumped to their deaths. This sanctification of a violent death helped to encourage and to justify a violent life. Being encouraged to die by the sword usually means that you will live by the sword, an important requirement in a warrior society like that of the Vikings, which required courage, boldness, and a proper fighting spirit for its survival (Colt, 1991, pp. 134-135).

The best term to describe the attitudes of inhabitants of the Greco-Roman world toward suicide would be ambivalence or even confusion. Greek culture condemned suicide, but Greeks were not particularly intolerant of it. Homer himself regarded suicide, if it were done voluntarily and for good reasons, as both natural and heroic (Droge & Tabor, 1992, pp. 17-18). Whatever disapproval of suicide existed in the ancient world was almost entirely dependent on the motives for the suicide. Self-killings done out of honor, devotion, or duty, to avoid dishonor, or as a response to unbearable grief were likely to be viewed as good and noble deaths. Only when self-killings evidenced signs of weakness, cowardice, or irresponsibility were they viewed as disgraceful. To the Romans, the reasons that one died were of fundamental importance in how the death was judged (Plass, 1995, p. 83). Suicides to avoid dishonor or as a response to the unbearable pain of the death of a loved one or of a love affair gone sour were not considered at all unreasonable (Farberow, 1988, pp. xi-xiii).

Christians were ultimately responsible for changing the ambivalence of the Greco-Roman world into intolerance and condemnation. In the beginning, Christians were not particularly intolerant of suicide. The Apostles did not moralize against it, and neither the Old nor the New Testament directly forbade it. In fact, in the early days of Christianity, suicide was viewed as a way to achieve martyrdom and to enter the kingdom of heaven. Everybody who wanted to get to heaven (and who really didn't?) could get there in a direct and expeditious way by committing suicide. The icing on the cake was that church members took care of the surviving family members of someone who committed suicide, which significantly eased the conscience of anyone who was planning to self-destruct (Farberow, 1988, p. xiii). Suicide evolved into a too-attractive option for Christians, so the church worked to harden attitudes toward it and to redefine it as an offense against God, society, and self.

The moral justification for the condemnation of suicide was found in the sixth commandment, "Thou shalt not kill." Life was declared to be a gift from God, and it was asserted that only God could determine when life was to end. People who took their own lives knew that they would be harshly dealt with in the hereafter. The severe sanction attached to suicide—eternal damnation—gave the ban a supreme moral authority and increased its potential to deter suicide (Barry, 1994, pp. 91-163). Though most Christians still wanted to get into heaven and viewed life on earth as transitory and unimportant, the surest route to paradise had been taken away. However, even here, the prohibitions against suicide were not absolute. When done for the right reasons—as a response to a loss of faith, as a response to personally disgraceful behavior, as a response to being captured by an enemy, or as a response to unbearable torture—suicide was an honorable death (Farberow, 1977).

The Victorians tended to sensationalize murder and to dread and to fear suicide (Gates, 1988). The suicide was treated as the lowest of the low. Surviving family members of a person who had committed suicide did all that they could to hide it because suicide was illegal, immoral, and shameful. The Crown could take all the property of a suicide, and the only way to avert this catastrophe was to convince authorities that the deceased was actually insane. Clerics were enlisted to aid in the conspiracy of silence because people who had killed themselves were not allowed to be buried on church property (the ban ended in the 1880s) (Gates, 1988, pp. 38-39). It was customary for the corpse of a suicide to be buried at a crossroads with a stake through its heart and a heavy stone over its face so that it could not return to harm the living. The last official record of a degradation of the corpse of a suicide in England was made in 1823. But the degradation did not stop: Bodies of suicides who had been poor during their lives or that were unclaimed were sent to medical schools for dissection and study for 50 years after that (Alvarez, 1972, pp. 46-47).

During the 20th century, attitudes toward suicide continued to fluctuate. Some tolerance could be found among the educated, but the Christian church remained steadfast in its opposition. Industrialization made it possible for some people to attain great prosperity but also made it possible for some people to become dismal failures. Poverty was equated with badness, and prosperity was associated with goodness. The anomie produced by a drastic and dramatic change in one's economic standing, especially from prosperity to poverty, was conducive to suicide (Lehmann, 1995, p. 917). The growth of a middle class, encouraged by the social changes of the Industrial Revolution, meant that more families had an important stake in conformity and in maintaining their status and prominence in the community. A suicide caused great pain for everyone in the family, and it reflected badly on the entire unit. The stigma of suicide was only partly lessened by the predilection of authorities to define suicide as an irrational act of an individual who was suffering from severe psychological problems.

Suicide in Different Places

People in non-Western, nonindustrial societies tend to have a generalized horror of the wandering spirits of dead people anyway, but the fear is all that much greater if the deceased died violently (Colt, 1991, p. 132). Whereas the spirit of a murdered individual is believed to have it in for only his or her killer, the spirit of a suicide is more likely—or so it is believed—to blame the entire group and to be bent on revenge of the worst kind toward the living (Fedden, 1938/1980, p. 35). For his or her part, a person who commits suicide has shown an alarming indifference to others. He or she probably has put surviving family members, friends, and acquaintances in the position of bearing the pain and shouldering the additional and difficult responsibilities that the decedent was unwilling or unable to manage. It is not hard to understand how some of the anger, regret, and sorrow that the living experience in response to the suicide of one of their own could get translated into a generalized fear of the deceased.

Some groups go to great efforts both to placate and to immobilize the spirit of a suicide to keep it from causing any mischief among the living. It was customary among a central African tribe called the Baganda to take the corpse of a suicide to a crossroads and to burn it. When females passed the spot of cremation, they would hurl grass or sticks on it as a way to keep the spirit from entering their bodies and being reborn. The Bannaus of Cambodia took the corpses of suicides far away to a corner of the forest to be buried, and the corpses of suicides in Dahomey were abandoned in the fields so that they would be eaten by wild animals. The Alabama Indians unceremoniously pitched the corpses of suicides into a river, and the Wajagga of East Africa would hang a goat in the same noose that the human used to commit suicide, hoping that it would placate the deceased's wandering spirit. In cultures across the world, a suicide's spirit is blamed for a host of catastrophes such as diseases, plagues, storms, famines, droughts, and bad harvests (Colt, 1991, p. 132).

One non-Western group with a high rate of suicide is the Netsilik (Balikci, 1989). Like the Vikings, the Netsilik believed that the souls of individuals who died violently (and suicide qualifies) went to a paradise, one of the three afterworlds recognized by Eskimos. However, this is not the entire story because if it were, practically everybody would have committed suicide to get to paradise. A more important factor was the deterioration of interpersonal relationships and the disappearance of traditional sources of mutual help and support (Balikci, 1989, p. 168). Families had left the region or been torn apart by the departure of family members, and kinship bonds had become more tenuous. The availability of rifles in the 1920s made it easier for a lone hunter to kill a caribou or seal without the cooperation of others, but the new technology produced further isolation from a traditional source of support and security. Men were very jealous over their wives, and considerable tension and insecurity characterized the marital relationship.

When a Netsilik suffered a personal misfortune—sickness, injury, loss of a loved or valuable friend, depression, despair—too few people were available to offer relief. Suicide was culturally defined as a quick road to a better existence, and people in the community rarely had the interest or inclination to stop a suicide or to help a suicidal individual find reasons to live. In the usual case, a desperate individual faced an uncertain future without the effective support of friends and family (Bergman & Dunn, 1990, p. 14). Males killed themselves more often than females. Most individuals hanged themselves using animal skins; about one third used guns, usually fired into the head; and a small number drowned themselves or suffocated themselves. The principal reason for a suicide was an injury to or a loss of a loved one, followed by personal problems such as injury, illness, or marital discord.

A breakdown in cohesion, intimacy, and traditional sources of group support seems to be conducive to an increase in suicide, and these disrupting forces can occur anywhere. The rate of suicide can increase significantly as a response to stresses and strains caused by outside intervention (e.g., the slave trade in Africa or the confinement of Native Americans on U.S. reservations) or by migration to a new territory. Different ethnic and racial groups adjust to new cultures and to new groups of people in different ways. The accommodation and resistance to patterns of prejudice and discrimination in a host culture can cause suicide among members of an oppressed or exploited minority group (Group for the Advancement of Psychiatry, 1989, p. 101). It is also true, however, that claims about high rates of suicide in some out-group may reflect ethnocentrism and faulty reasoning among members of the defining group (Webb & Willard, 1975, p. 17).

An exploration of the attitudes in the African American community toward self-destruction shows how subcultural meanings and suicide are related (Early, 1992; Early & Akers, 1993). Other-directed violence, alcohol, and drugs are understandable responses, albeit unfortunate, to the endemic racism of U.S. society. However, these same experiences have promoted the development of a culture of resilience to factors that would otherwise result in an increase in suicide. "Suicide is the ultimate, unforgivable, and unredeemable offense. Abuse of alcohol and drugs, participation in the drug trade, and even violence, although strongly condemned, are not viewed as inexplicable, unforgivable, or unredeemable" (Early & Akers, 1993, p. 291). These social meanings of suicide make suicide not only unacceptable but practically unthinkable for most members of the African American community (Early & Akers, 1993, p. 292).

Obligatory Suicide

One of Durkheim's (1897/1951) more provocative observations is that suicide can occur when social integration is high and the group becomes more important than any individual in it (p. 217). An example of this type—Durkheim called it "altruistic suicide"—is the death of a soldier who jumps voluntarily on a live grenade to protect other members of the platoon. In some places, altruistic suicide ceases to be optional, and certain individuals are put under great pressure to kill themselves or to be scorned by other members of the group (Durkheim, 1897/1951, pp. 221-222). These obligatory suicides are interesting in their own right, but they also show the importance of definitions and social constructions. Is not an obligatory suicide actually a murder?

A good example of obligatory suicide is the custom of *suttee* or *sati,* in which a widow casts herself upon her deceased husband's funeral pyre, forfeiting her own life. It was named in honor of a heroine from Hindu mythology who committed suicide to prove her wifely devotion. The options confronting a widowed woman were very few, and none of them were particularly attractive from her standpoint. She could choose not to commit suicide and lead a miserable and unrewarding life for her remaining days, or she could leap onto the flames and receive the honor and veneration of all who knew her (Roy, 1987, p. 75). A widowed woman in Hindu society had no identity apart from her husband (Sharma, 1988). Marriage to a second husband was rare, and it usually meant that the woman was rejected by her own family. Not only did people not feel sorry for her, they might even blame her for her husband's death (Saxena, 1975, pp. 66-76). The rewards of committing suicide, coupled with the punishments for refusing, were sufficient to propel most widowed women in the direction of self-immolation (Rao, 1983, p. 212). Even those women who were reluctant to follow their husbands into death were easy enough to manipulate. The average widow was young—in some cases no older than 10—grief stricken, and suggestible. To guarantee that she joined her husband, the bereaved widow might have been given opium, marijuana, or some other drug that would make it even more likely that she would do what she was told without complaint (Datta, 1988, pp. 207-221).

The custom of *suttee* was not restricted to India, being found among the Greeks, Egyptians, Germans, and Chinese. If the deceased was well-to-do, he would be followed in death by several wives, slaves, and servants. Once in a while, other people felt the pinch of bereavement, and they too would commit suicide to be with the decedent: mothers, sisters, sisters-in-law, ministers, and nurses (Thakur, 1963, pp. 142-143).

Death before dishonor was the rule to live by in the Orient, at least among samurai warriors. The Japanese institutionalized suicide in what was called *hara-kiri* (belly-slitting) or the more elegant term of *seppuku*. For centuries, it was a privilege reserved by law for the samurai; it was not outlawed until 1868. A ritualized suicide such as *hara-kiri* was the only way for a dishonored warrior to reclaim his honor (while forfeiting his life). Sometimes it was done to display allegiance to a fallen leader or to avoid the dishonor of a public execution (Evans & Farberow, 1988, pp. 175-176). *Hara-kiri* was usually done according to a precise ritual, in special surroundings, in front of a group of invited guests:

> The samurai knelt on a white hassock on a white-edged *tatami* (reed mat), facing a small white table on which lay a short sword with its blade wrapped in white paper. Taking the blade by its middle in his right hand, the samurai made an incision in the left side of his abdomen, drew the blade to the right, and then made an upward cut. It was meritorious then to make another incision in the chest and a downward cut, allowing the entrails to spill out. (O'Neill, 1981, pp. 12-13)

Because dying could take several hours, a friend was usually close at hand to end the warrior's suffering by beheading him with a sword. The samurai's wife might choose to die with her husband by stabbing herself in the throat with a short dagger known as a *jigai,* a wedding present given to her explicitly for this purpose.

The kamikaze pilots of World War II were the heirs apparent to the samurai warriors. In a last-ditch effort to win the war, Japanese pilots flew their planes into the ships of their enemy, killing themselves in the process (O'Neill, 1981, p. 140). The young men who volunteered for the suicide missions were motivated by strong feelings of loyalty and patriotism. Doggedly pursuing victory, regardless of the chances of winning, was a Japanese tradition. The greater the prospects of defeat, the more committed the Japanese became to continuing the fight. The emperor was divine, and so was his word. Any war that he authorized automatically became a holy war, and losing it was unacceptable. Japanese soldiers believed that if they died heroically, they would become gods themselves and join the guardian spirits of the country at Yasukuni Shrine of Kudan Hill. The kamikazes were willing to die for the sake of their families, their relatives, and their emperor in the fervent belief that by sacrificing their lives they would make a difference in the world (Naito, 1989). Put simply, a suicide mission was viewed as a glorious way to die (Spurr, 1981).

Revenge Suicide

Why would anyone kill himself or herself to get revenge on someone else? Durkheim helps us again. He thought that fatalism—a condition of overregulation and excessive discipline—produced feelings of despair and hopelessness in the regulated individuals to such an extent that death was the only way out for them (Durkheim, 1897/1951, p. 276). What Durkheim did not seem to imagine was that suicide could be carried out to hurt or to get even with the source of the misery and the pain. Fedden (1938/1980, p. 45) offers a bizarre example of a revenge suicide. A Frenchman of the 19th century had been betrayed by his mistress. He was so upset over her unfaithfulness that he killed himself. He had instructed his servant to get his corpse and fashion a candle from its fat, light it, and carry it to his ex-lover. It was accompanied by a letter in the suicide's own hand that stated that as he had burned for her in life, he now burned for her in death. She read these words by the light given off by his burning flesh! Douglas (1967) claimed that committing a suicide as a way to get revenge on someone and to blame him or her (or them) for the death is one of the distinctive meanings of suicide in the Western world (p. 311). Douglas was too restrictive; revenge suicide is found throughout the world.

In many non-Western societies, suicide is viewed as a good way for some people to get back at others who have wronged them. Malinowski (1926) described revenge suicides among the Trobriand Islanders. They could shinny up a palm tree, loudly name their tormenters, and then jump off the tree and fall to their deaths. The rule was a suicide for a suicide, so the marked individual had to respond in kind (or take some effort to placate the spirit of the decedent or to compensate the decedent's family). The Tshi-speaking people of Africa believe that one person can drive another person to suicide. If a person commits suicide and blames the death on another, the named party is required by native custom to commit suicide also (Counts, 1990, p. 95).

Revenge suicide is socially patterned and is designed to convey a great deal of meaning to interested observers (Counts, 1990, p. 95). A culture must recognize the possibility that a living individual can be held responsible for the suicidal death. This requires some ethic of shared or collective responsibility, where, for example, a husband can be blamed for his wife's misery and subsequent suicide. A culture must also provide definite mechanisms whereby the individual who kills himself or herself can clearly show who is responsible for the suicide and make the living willing to punish the responsible party (Counts, 1990, p. 95). In some cases, the living come to believe that if they fail to avenge the death, they themselves will be haunted by the angry spirit of the deceased.

A revenge suicide is most likely to be successful and the living are most likely to administer a negative sanction to the guilty party if the individual commits the suicide in a public place where the body definitely will be found, forewarns others of the upcoming suicide, and clearly identifies whomever is being held responsible for the death. Revenge suicides are usually done by powerless individuals, often females, who have been used and abused by people in their lives. The only way for these individuals to retaliate against those people who have hurt them is for them to kill themselves in a public way. Their hope is that they will shame their tormentors enough that these people will want to kill themselves for all the injury that they have caused (Counts, 1990, p. 99).

Assisted Suicide—The Controversy Rages

A pitched battle rages in the United States between proponents of assisted suicide and its critics. The crux of the matter concerns the very issues that we have examined throughout this chapter. How is death viewed, and what factors determine the way a death is classified and evaluated? An important event in the ongoing dispute over assisted suicide was the involvement of Dr. Jack Kevorkian in the death of Janet Adkins.

Ms. Adkins was a happily married, 54-year-old mother of three grown sons who was afflicted with Alzheimer's disease. She strongly wanted to end her own life before the degenerative disease ruined her ability to take care of herself. She had tried to get help from physicians in Portland, Oregon—the city where she lived—but all of those whom she asked refused to help her die. Janet Adkins learned of Kevorkian in 1989, and she even saw him explain his procedures for assisted suicide on a national television show. Her objective was to die with dignity and without pain, and Kevorkian seemed to be the only person who could help her. She sent him records of her medical history, and he wanted to know more. She and her husband flew from Portland, Oregon, to Royal Oak, Michigan. After talking to her, Kevorkian was convinced that she was rational and suffering from an irreversible physical condition, so he decided to help her die. All his efforts to find a place to "host" the suicide that was befitting to its solemnity were unsuccessful. Her final exit came on June 4, 1990, in a rather unceremonious setting: the back of Kevorkian's Volkswagen van, parked in a Michigan campsite. When she was ready, Janet Adkins pushed a button, and Kevorkian's suicide machine did the rest: One drug put her to sleep, and another drug killed her. Ms. Adkin's husband awaited news of her death at a nearby hotel because his wife would not allow him to be present at the actual death scene (Humphry, 1991, pp. 131-141).

Assisted suicide may mean little more than the discontinuation of life support efforts for terminally ill individuals. In these cases—sometimes called "death with dignity"—the plug is pulled, and no further efforts are made to keep an individual alive. However, no active efforts are taken to stop an individual's life. One case that galvanized attitudes about this type of assisted death was the death of Karen Quinlan (Marcus, 1996, pp. 171-172). Ms. Quinlan lapsed into unconsciousness in April 1975 after she attended a party and drank too much alcohol and took too many tranquilizers. Her breathing stopped long enough that it caused permanent brain damage. The physicians who examined her all agreed that she would always be in a vegetative state. Her parents were assured by a Roman Catholic priest that it was not immoral for them to discontinue life support efforts when no possibility of improvement existed. However, physicians and hospital administrators refused to allow the respirator to be turned off. The parents went to court, asking that Karen be allowed to die with grace and dignity, but the trial judge refused to grant their request. In March 1976, the New Jersey Supreme Court reversed the lower court's ruling. Karen had a right to refuse treatment but because she was unable to make the determination herself, the decision fell to her parents. In 1976, the respirator was turned off, she started breathing on her own, and she survived for another 9 years.

How assisted suicide is framed is critical in determining the levels of support for it. Successful presentation of an assisted suicide as self-deliverance from an irreversible terminal illness, a dignified death, or a simple medical procedure makes it much more likely that it will be viewed favorably than if it is called suicide, lethal injection, or euthanasia (Worsnop, 1995, p. 398). Support rises if safeguards exist—clear and consistent ones—that make assisted suicides fully voluntary and fully informed and if people are truly suffering from an incurable, irreversible disease. Because of the secrecy and mystery that now surround physician-assisted suicide, a greater possibility exists that suicide could be carried out by an individual who was tired of living or depressed, rather than by a person who was terminally ill. Ms. Adkins had played tennis with one of her sons a few days before her death (although she was too mentally incapacitated to keep the score). Some critics of assisted suicide are fearful that a right to die could easily evolve into an obligation to die (Hendin, 1995a). How do we know or recognize an incurable disease when we see one? Once active euthanasia is firmly in place, it may become more likely that it will be called into service to get rid of people who are not terminally ill and who have not given their full and informed consent.

On December 3, 1990, Kevorkian was charged with the murder of Janet Adkins. A few weeks later, all charges were dismissed because of lack of evidence that Kevorkian had planned her death and intentionally caused it. An additional factor was that Michigan had no law directly forbidding an individual from helping another to commit suicide. In 1992, the Michigan legislature did pass a bill that banned assisted suicide, and the Michigan Supreme Court decided that assisted suicides could be prosecuted as felonies.

However, Michigan prosecutors still found it difficult to get Kevorkian convicted of a crime. Five times they tried, and five times they failed. The trials ended either in acquittals or, in one instance, a mistrial. However, on April 13, 1999, that changed. Kevorkian was sentenced to 10 to 25 years in prison for the murder (second degree) of 52-year-old Thomas Wouk, who suffered from Lou Gehrig's disease (amyotrophic lateral sclerosis). Kevorkian had not only supplied the drugs used to kill Wouk, he also had injected the drugs himself into the man. What's more, he had recorded the whole thing on film. With all this evidence, prosecutors were able to charge Kevorkian with murder and make it stick.

Disagreements over the deviancy of suicide have important consequences for how deviants are viewed. Prosecutors described Kevorkian as "a medical hit man," while Kevorkian portrayed himself in a different and more positive way. He claimed that what he did was no different from the actions of Martin Luther King Jr., Rosa Parks, or other civil rights activists who deliberately broke laws because these laws were immoral and unjust. Kevorkian compared himself to an executioner instead of to a murderer. One thing is clear: In order for assisted suicide to receive greater acceptance, safeguards must be developed and implemented (Cosculluela, 1995). The suicide must be fully intended; the individual must be fully informed; and, most important, the self-inflicted death must be the individual's last remaining option.

Summary and Conclusions

In this chapter, we have examined information dealing with the relativity and ambiguity of suicide. At some times, in some places, suicide has been met with admiration and widespread social support. At other times, in other places, suicide has been condemned and treated as evidence of immorality, disease, or antisocial tendencies.

Our examination showed us that suicides are part of social structure and reflect different patterns of social organization, ethnic and racial concentrations, and cultural beliefs. These social factors can have direct and important influences on suicide separate from an individual's levels of hopelessness, helplessness, or depression. Death is symbolized in many different ways, and death of the physical body does not always denote death of the soul or spirit. Historical and cross-cultural variations in the symbolization of suicide show that contradictions exist regarding death and what it means. Before individuals frame some death as a suicide instead of a murder, an accident, or a natural death, they must always put ambiguous bits of information together into some consistent whole that will satisfactorily explain what really happened, why, and who is responsible.

References

Alvarez, A. (1972). *The savage god: A study of suicide*. New York: Random House.
American Psychiatric Association. (1994). *Diagnostic and statistical manual of mental disorders* (4th ed.). Washington, DC: Author.
Andrade, M. C. B. (1996). Sexual selection for male sacrifice in the Australian redback spider. *Science, 271,* 70-71.
Balikci, A. (1989). *The Netsilik Eskimo*. Prospect Heights, IL: Waveland.
Barry, R. (1994). *Breaking the thread of life: On rational suicide*. New Brunswick, NJ: Transaction.
Bergman, B., & Dunn, J. (1990, June 4). Northern agony. *Maclean's,* p. 14.
Calhoun, L., & Selby, J. W. (1990). The social aftermath of a suicide in the family: Some empirical findings. In D. Lester (Ed.), *Current concepts of suicide* (pp. 214-224). Philadelphia: Charles.
Canetto, S. S., & Lester, D. (1995). Women and suicidal behavior: Issues and dilemmas. In S. S. Canetto & D. Lester (Eds.), *Women and suicidal behavior* (pp. 3-8). New York: Springer.
Colt, G. H. (1991). *The enigma of suicide*. New York: Summit.
Cosculluela, V. (1995). *The ethics of suicide*. New York: Garland.
Counts, D. A. (1990). Abused women and revenge suicide: Anthropological contributions to understanding suicide. In D. Lester (Ed.), *Current concepts of suicide* (pp. 95-106). Philadelphia: Charles.

Datta, V. N. (1988). *Sati: A historical, social and philosophical enquiry into the Hindu rite of widow burning*. Riverdale, MD: Riverdale.

deCatanzaro, D. (1981). *Suicide and self-damaging behavior: A sociobiological perspective*. New York: Academic Press.

Diekstra, R., & Garnefski, N. (1995). On the nature, magnitude, and causality of suicidal behaviors: An international perspective. In M. Silverman & R. Maris (Eds.), *Suicide prevention toward the year 2000* (pp. 36-57). New York: Guilford.

Douglas, J. (1967). *The social meanings of suicide*. Princeton, NJ: Princeton University Press.

Droge, A. J., & Tabor, J. D. (1992). *A noble death: Suicide and martyrdom among Christians and Jews in antiquity*. San Francisco: HarperSanFrancisco.

Durkheim, E. (1951). *Suicide: A study in sociology* (J. Spaulding & G. Simpson, Trans.). New York: Free Press. (Original work published 1897)

Early, K. (1992). *Religion and suicide in the African-American community*. Westport, CT: Greenwood.

Early, K., & Akers, R. (1993). "It's a white thing": An exploration of beliefs about suicide in the African-American community. *Deviant Behavior, 14,* 277-296.

Evans, G., & Farberow, N. (Eds.). (1988). *The encyclopedia of suicide*. New York: Facts on File.

Farberow, N. (1975). Cultural history of suicide. In N. Farberow (Ed.), *Suicide in different cultures* (pp. 1-15). Baltimore: University Park Press.

Farberow, N. (1977). Suicide. In E. Sagarin & F. Montenino (Eds.), *Deviants: Voluntary actors in a hostile world* (pp. 503-570). Morristown, NJ: General Learning Press.

Farberow, N. (1988). Introduction: The history of suicide. In G. Evans & N. Farberow (Eds.), *The encyclopedia of suicide* (pp. vii-xxvii). New York: Facts on File.

Fedden, H. R. (1980). *Suicide: A social and historical study*. New York: Arno. (Original work published 1938)

Gates, B. T. (1988). *Victorian suicide: Mad crimes and sad histories*. Princeton, NJ: Princeton University Press.

Goffman, E. (1974). *Frame analysis: An essay on the organization of experience*. New York: Harper & Row.

Group for the Advancement of Psychiatry. (1989). *Suicide and ethnicity in the United States*. New York: Brunner/Mazel.

Hazelwood, R., Dietz, P. E., & Burgess, A. W. (1982). Sexual fatalities: Behavioral reconstruction in equivocal cases. *Journal of Forensic Science, 27,* 763-773.

Hendin, H. (1995a). Assisted suicide, euthanasia, and suicide prevention: The implications of the Dutch experience. In M. Silverman & R. Maris (Eds.), *Suicide prevention: Toward the year 2000* (pp. 193-204). New York: Guilford.

Hendin, H. (1995b). *Suicide in America* (new, expanded ed.). New York: Norton.

Henry, J., & Short, A. (1954). *Suicide and homicide*. New York: Free Press.

Humphry, D. (1991). *Final exit: The practicalities of self-deliverance and assisted suicide for the dying*. Eugene, OR: Hemlock Society.

Jacobs, J. (1970). The use of religion in constructing the moral justification of suicide. In J. D. Douglas (Ed.), *Deviance and respectability: The social construction of moral meanings* (pp. 229-251). New York: Basic Books.

Lehmann, J. (1995). Durkheim's theories of deviance and suicide: A feminist reconsideration. *American Journal of Sociology, 100,* 904-930.

Littlewood, R., & Lipsedge, M. (1987). The butterfly and the serpent: Culture, psychopathology and biomedicine. *Culture, Medicine and Psychiatry, 11,* 289-335.

Lowery, S., & Wetli, C. (1982). Sexual asphyxia: A neglected area of study. *Deviant Behavior, 4,* 19-39.

Malinowski, B. (1926). *Crime and custom in savage society.* London: Routledge & Kegan Paul.

Marcus, E. (1996). *Why suicide? Answers to 200 of the most frequently asked questions about suicide, attempted suicide, and assisted suicide.* New York: HarperCollins.

Naito, H. (1989). *Thunder gods: The kamikaze pilots tell their story.* Tokyo: Kodansha.

O'Neill, R. (1981). *Suicide squads.* London: Salamander.

Papke, D. R. (1987). *Framing the criminal.* Hamden, CT: Archon.

Plass, P. (1995). *The game of death in ancient Rome: Arena sport and political suicide.* Madison: University of Wisconsin Press.

Pope, W. (1976). *Durkheim's* Suicide: *A classic analyzed.* Chicago: University of Chicago Press.

Poppel, F. van, & Day, L. H. (1996). A test of Durkheim's theory of suicide—without committing the ecological fallacy. *American Sociological Review, 61,* 500-507.

Rao, A. V. (1983). India. In L. Headley (Ed.), *Suicide in Asia and the Near East* (pp. 210-237). Berkeley: University of California Press.

Resnik, H. L. P. (1972). Erotized repetitive hangings: A form of self-destructive behavior. *American Journal of Psychotherapy, 26,* 4-21.

Rodgers, L. (1995). Prison suicide: Suggestions from phenomenology. *Deviant Behavior, 16,* 113-126.

Roy, B. B. (1987). *Socioeconomic impact of sati in Bengal and the role of Raja Rammohun Roy.* Calcutta: Naya Prokash.

Saxena, R. K. (1975). *Social reforms: Infanticide and sati.* New Delhi: Trimurti.

Sharma, A. (1988). *Sati: Historical and phenomenological essays.* New Delhi: Motilala Banarsidass.

Shneidman, E. (1985). *Definition of suicide.* New York: John Wiley.

Spurr, R. (1981). *A glorious way to die.* New York: Newmarket.

Taylor, S. (1990). Suicide, Durkheim, and sociology. In D. Lester (Ed.), *Current concepts of suicide* (pp. 225-236). Philadelphia: Charles.

Thakur, U. (1963). *The history of suicide in India: An introduction.* New Delhi: Munshi Ram Manohar Lal Oriental Publishers.

Unnithan, N. P., Huff-Corzine, L., Corzine, J., & Whitt, H. P. (1994). *The currents of lethal violence: An integrated model of suicide and homicide.* New York: State University of New York Press.

Uva, J. (1995). Review: Autoerotic asphyxiation in the United States. *Journal of Forensic Sciences, 40,* 574-581.

Webb, J., & Willard, W. (1975). Six American Indian patterns of suicide. In N. Farberow (Ed.), *Suicide in different cultures* (pp. 17-33). Baltimore: University Park Press.

Worsnop, R. (1995, May 5). Assisted suicide controversy. *CQ Researcher, 5,* 393-416.

Wright, R. K., & Davis, J. (1976). Homicidal hanging masquerading as sexual asphyxia. *Journal of Forensic Sciences, 21,* 387-389.

7

Other People's Belongings

Introduction: Take the Money and Run

Pretend that you're walking down the main street of Metropolis, USA, a news-paper under your arm. A man walks up and borrows your paper. He spreads it over the top of a rusty oil drum, takes out three playing cards (two red aces and a black queen), and throws them face down on the newspaper. He picks them up, shows the location of the queen, and throws them down again. A crowd gathers, and the man gives the following spiel:

> If your eye is faster than my hand and you find the queen, I pay you the same amount you bet. I'll accept bets of five, ten, twenty, fifty or a hundred that you can't find the lady. Remember, if you don't speculate, you can't accumulate. Money in hand or no bet. Let's go. (Scarne, 1974, p. 620)

He throws the cards, and the odds seem good to you—one chance in three of winning as much as you bet. So you bet $10 you can find the queen, but you lose. The man throws once more. Again you think that you know where the queen is, so you bet another $10. Once again you lose. He throws them one more time, and you decide to let someone else take a chance. That player wins $15. You decide to try your luck again. You reach for the card that you think is the queen with your $10 bet in hand, but you lose again. However, good fortune is about to come your way, or so you think.

One of the players crimps or bends slightly the corner of the queen while the card thrower is distracted. The "tosser" picks up the cards and then throws them down again. The player who bent the corner bets $50 that he can find the queen, selects the bent card, and wins with no trouble. The cards are thrown again. A different player bets on the bent card and wins. The card tosser then announces that this throw is the last:

> I'll accept how ever much you have remaining in your bankroll—$50, $100, $500—it makes no difference. But after I throw the cards if anyone touches a card before I stop taking bets, all bets are off. Okay, here goes. (Scarne, 1974, p. 623)

The cards are thrown, and the card with the bent corner falls right in the middle. You and everybody else in the crowd bets on it, and the tosser tells someone to turn it over. All the players are really feeling confident, but when the card is turned over, it is an ace, not the queen as everyone thought. While everybody stands there stunned, the card tosser folds the newspaper with the cards inside (your newspaper) and beats a hasty retreat. As you walk away, you say to yourself you have just been beaten by an expert with cards and lady luck. However, you are only partially correct. It was actually the social organization of thievery that did you in.

You have just been fleeced by one of the oldest American swindles, three-card monte. It was played on the riverboats that traveled on the Ohio and Mississippi Rivers, and it was also played throughout the West in the 1850s (Scarne, 1974, p. 619). In fact, it is likely that three-card monte was the forerunner of *all* contemporary confidence games (Maurer, 1974). A tosser uses simple sleight of hand to substitute an ace for the queen as the cards are being released. The switch is virtually undetectable when done well, and it would be better for a player to pick randomly than to try to follow the flight of the queen with the eye. As skilled as the tosser may be, if he were all by himself, he could not guarantee winning every time, nor could he protect himself from unhappy or disgruntled players. To guarantee the greatest returns with the least amount of risk and effort, thieves organize themselves into a mob.

The important role played by the secret collusion between members of the mob can be seen in the "bent-corner ruse." In our hypothetical case, when you bet on the card with the bent corner, you lost because the card turned out to be an ace instead of the queen. Think how effective that is. You have been consistently losing. Finally, a sure thing arises. People who bet on the bent corner have actually won. Your chance to get something for nothing and to recoup your losses has arrived. The card tosser announces that the game will end after one more throw. It's your last chance, so the urge to bet is practically irresistible. However, you still lose, and here's why. A member of the mob (known as a shill) was the one who bent the corner, and members of the mob are the only ones who won by betting on the bent card. Before the last throw, in the process of gathering the cards, the card tosser secretly unbends the corner of the queen and secretly bends a corner of an ace. Everybody who bets on the bent card loses.

Three-card monte is a simple scam, engineered by people without a great deal of skill. The dishonesty of the game and its profitability cannot be understood apart from the web of relationships and covert understandings among members of the gang. These relationships increase this group's ability to take other people's belongings. Beyond that, these relationships maximize the chances of escape. If the scam is done well enough, the players don't even know that they've been the victims of a theft; they believe that they were simply unlucky and remain confident that their luck will be better some other time.

Power Corrupts, and Absolute
Power Corrupts Absolutely

We cannot separate our understanding of the relativity of deviance from our understanding of power—who has it, how much they have, and how power affects the ebb and flow of social life (Davis, 1980). Power over others is based partly on an individual's knowledge and skills. The card tosser in three-card monte had power partly because the thief knew how to manipulate the cards in such a way that it increased the odds of winning. The tosser also presented and maintained a facade and kept cool and calm throughout the swindle. Not all people can do this, and certain scams and thefts require greater skills and knowledge than others. It is more difficult, for example, to maintain one's composure during a high-stakes poker game while cheating others—bottom-dealing, holding out cards, stacking a deck, undoing the cut—than it is to master the throw used in the game of three-card monte (Prus, 1977).

A enduring and important source of power comes from an individual's group affiliations, social networks, and social statuses that enable him or her to do things that affect the lives of others. Conan the Barbarian has a certain amount of power because he is big and strong; the chief executive officer of General Motors, however, has infinitely more power even though the CEO may have small stature and be unable to lift a feather. When powerful people with interests in common unite to accomplish their objectives and establish ties to other powerful groups, they can move mountains. They can also use their powers in a harmful or malicious way.

Powerful people have greater opportunities to commit deviance, and they often experience a more intense sense of subjective deprivation. These two factors coalesce to make it more likely that the powerful will engage in harmful acts of deviance more frequently than the powerless (Thio, 1995). The powerful may deny that anyone was harmed by what they did, blame their deviance on some external factor over which they have no control, or convince themselves that they deserve the rewards that their deviance brought them (Coleman, 1987). They may use their resources to mask their untoward behaviors and to direct attention to the deviance of the powerless. The powerful benefit from a "cloak of immunity" (Box, 1983, p. 99) or "shield of elitist invisibility" (Simon, 1996, p. 38), such that their rule breaking either is not uncovered or, if uncovered, is accorded only a mild penalty. "The process of law enforcement, in its broadest possible interpretation, operates in such a way as to *conceal* crimes of the powerful against the powerless, but to *reveal* and *exaggerate* crimes of the powerless against 'everyone' " (Box, 1983, p. 5). The ruin caused by powerful organizations has a good chance of being peddled as an accident, as an isolated event (and therefore less harmful), or as the result of the antisocial behaviors of just a few "bad apples" or "mavericks" (Poveda, 1994).

The Carrier's Case: Power, Signification, and Theft

The transformation of the "taking of other people's belongings" into the crime of theft is an important event. It reconstitutes the behavior: What was once seen as business as usual or perhaps individual nastiness comes to be seen as a social problem with broader implications. The perpetrator now has a different status—criminal—and can be formally and forcefully sanctioned with the full weight of government. According to Quinney (1970), crime is a definition of behavior that is conferred on people and behaviors by agents of law (legislators, police, prosecutors, judges) in politically organized society. Crime is created by those authorities who are able to formulate and to apply criminal definitions (p. 15).

Though it may seem natural to want to own things and to protect them from others, laws protecting property as we know them now were nonexistent before the 15th century. Their creation had practically nothing to do with a belief in the divine right of all persons to be safe and secure in their possessions. The laws regarding theft—the norms that transformed taking other people's belongings into a crime—were established by powerful individuals for the purpose of protecting their own interests. Any benefit that these laws may have had for others was purely coincidental.

The case that is responsible for the creation of the crime of theft occurred in England in 1473. It is known as the "carrier's case," and here is what happened. A man was hired to carry bales of cloth to Southampton. Instead of completing the job, he broke the bales open and took their contents for his own use. He was eventually caught and taken into a court of law, and the case was reviewed. The judges were in total agreement over the details of what the carrier had done. More important, they were in total agreement that what the carrier had done was entirely legal according to the prevailing laws. Because the carrier had been given the bales lawfully by his employer, nothing that he did after he took possession violated any existing law. Theft at the time was defined in terms of trespass, and nobody who had been given property by its owner could trespass against it. Legally, it was the owner who was held at fault for not employing a more trustworthy carrier (Hall, 1952, p. 5).

The judges were under pressure to craft the necessary legislation and to use the carrier as both an opportunity and a deterrent. So they decided that when the carrier broke open the bales, the property instantly reverted back to the owner. When the carrier absconded with the contents, it was a form of trespass after all and could be treated as a crime (Hall, 1952, p. 10). The judges found him guilty of theft. Curiously enough, if the carrier had somehow known what the judges were going to do, he could simply have taken the bales without breaking them open (but then the law would probably have been crafted differently). The judges portrayed the new law and the new understandings of "crime" and "criminal" that it established as just one more example of the strict application of legal precedent.

Why would the judges in the carrier's case break with established tradition and invent a new law? It was not really because they felt sorry for this particular owner or even because they wanted to encourage the carrier to walk the straight and narrow. To understand why a new law was crafted it is necessary to look at changing social conditions in England in the 15th century. Feudalism and its system of relationships based on sentiments of fidelity and obligation between serf and master were evaporating and being replaced by mercantilism and a middle class that owed its livelihood to commerce and trade. Any threat to the transportation of goods was a blow to this burgeoning economic system. King Edward IV, the ruling monarch at the time of the carrier's case, was highly supportive of trade, and he assiduously courted business interests. The owner of the property in the carrier's case was an Italian merchant who had been assured of safe passage for his cargo by King Edward IV himself. The bales contained wool or cloth, merchandise of great importance in England's budding economy.

The fact that the material was stolen while in transport to Southampton was even worse because it undermined business confidence in the security of a major English trading center and shipping port. Even worse, the goods were stolen by a professional carrier. Merchants were finding it more and more difficult to use their own servants to transport goods and were being forced to rely on carriers more and more. If carriers could not be trusted to deliver safely that which they were given, trade would not flourish as it should.

When the carrier stole the contents of the bales he was doing much more than depriving a solitary foreign merchant of his goods. He was threatening the economic interests of the textile industry, one of the most powerful and important businesses in all of England (Hall, 1952, p. 31). He was also challenging the interests of the Crown, both personally and as a warrantor of the safety of economic transactions in the British state. King Edward IV exerted a great deal of control over public officials, so the courts were likely to adopt any legislation that he wanted. What the new law of theft did, enforced by the courts, was to make taking property without consent of the owner a felony so that guilty parties were subject to severe penalties. During the centuries following the carrier's case, laws of theft were expanded to the point that thefts by cashiers and clerks eventually became illegal, as did the receiving of stolen property (Hall, 1952).

The law is a tool that powerful groups can use to give them superior moral as well as coercive power in conflicts with others (Chambliss & Seidman, 1982). Law is symbolic: It sets the broad parameters of correct and incorrect ways of acting in a society. Whether the law is actually followed is less important than that practically everybody knows that it is *supposed* to be followed. Law is also instrumental: It encourages people to follow the straight and narrow under the threat of unpleasant and unwanted penalties. The law of theft that was created in 1473 in Britain was a sop to powerful interests by powerful interests. It gave important individuals who felt victimized a way to protect themselves against the trespasses of others. It also provided a new understanding about proper and improper ways of doing business. Merchants' interests and self-serving behaviors were protected and reinforced, while the interests and self-serving behaviors of the carriers were singled out for control and correction. The carrier's case suggests that anything could be made illegal—and the new rules and sanctions justified as customary and fitting precedent—if the groups that create and enforce laws want this to happen. A corollary is that nothing will be illegal unless groups of people with adequate resources take the time and trouble to criminalize those things that they do not like.

Avarice, Insensibility, and the
Theft of a Neighborhood

The carrier in the carrier's case did something that is very predictable: He acted to feather his own nest instead of doing what was politically correct or morally just. This predilection to look out for oneself is not at all unusual. However, if people have power because of their social positions or interpersonal networks, it is easier for them to get what they want and then to portray their actions as reasonable and necessary. These efforts at legitimation help to free them from the kind of blame and derision that would befall thieves of lesser means.

A community in Detroit—once known as Poletown—traced its roots back to the 1880s. It started as a low-income, working-class, well-integrated, close-knit community whose residents had migrated from areas in Poland under Russian, Austrian, or German control. It developed into an area with strong community spirit, occupied by people who had a faith in the future. However, it all came to an end because Poletown, its residents, and much of what they owned—homes, businesses, hospitals, schools, churches, restaurants, laundromats—stood in the way of the interests of the privileged and the powerful. Here we have an example—unlike the carrier's case, in which a lowly person stole something from the powerful—of the powerful taking from others. An exploration of the pilfering of Poletown will help us to see what happens when the powerful flex their muscles and go after other people's belongings (Wylie, 1989).

In June of 1980, just before Ronald Reagan would be sent to the White House as president, the mayor of Detroit, Coleman Young, and the chairman of General Motors, Thomas Murphy, announced that a new Cadillac plant (the Hamtramack plant) would soon be built in Detroit on a site that would include a large portion of Poletown. The plant was packaged as a renovation project for Detroit that would create jobs (GM claimed that 6,000 workers would eventually be employed at the plant) and bring new prosperity to a city hit hard by economic recession and corporate flight. All seemed right with the world, and GM representatives and political officials apparently expected that everyone would be as pleased as they were with the prospects of a partnership between General Motors and the Detroit City Council.

The city council believed that it could iron out any wrinkles in the plan, pacify disgruntled Poletowners, and smoothly and quickly take care of business for General Motors. However, the best laid plans often go astray. Almost as soon as the plan was unveiled, it was criticized by those people who defined it as nothing more than a land grab to make one of the richest companies on earth even richer.

Members of Detroit's city council seemed befuddled to the point of amazement by the censure that they, General Motors, and the plan received. Some members of the city council claimed surprise and shock over the Poletowners' reaction to what they believed was one of the best things ever to happen to the city of Detroit. Others on the city council expressed anger over the vitriolic attacks to which they were subjected. The mayor of Detroit even went so far as to characterize attacking General Motors as equivalent to shooting Santa Claus (Wylie, 1989, p. 201).

It is hard for an outside observer to understand how anyone on the city council could reasonably believe that they would *not* be savagely attacked for what they proposed to do. Building of the plant was to require the taking (stealing?) of 465 acres from residents of Poletown, where their homes, businesses, churches, and schools—their memories and traditions—were located. Even if residents could convince themselves that the coming of new jobs to the area somehow justified the taking of most of what they held near and dear, a problem still existed: The proposed plant would probably not employ anywhere near what was projected because the plant was designed to be highly automated. Machines, rather than men and women, would do much of the work, and what new employees would be required would probably come from two older Cadillac plants that were scheduled to be closed by General Motors. This would mean job retention more than job creation, and it undercut the only legitimate reason—more jobs in Detroit—for the close alliance between the city council and GM.

General Motors was in a position where it could make demands without Detroit shutting the door. In fact, it seemed that GM held all the cards and called most of the shots. Because corporate spokespersons successfully steered attention to the jobs that would come to Detroit and away from the destruction and damage that would be done to community integrity and the profits that GM would make from the deal, GM's demands fell on receptive ears. General Motors demanded $350 million that actually belonged to local, state, and federal taxpayers to clinch the deal. General Motors also wanted the city to give it tax abatements; necessary permits (air, water, and waste); land rezoning; and city-funded roads, rail lines, highways, sewage removal facilities, and street lights (Wylie, 1989, p. 52). Detroit officials had to obtain title to Poletown's 1,400 homes, 144 businesses, two schools, a hospital, and 16 churches and destroy them within 10 months—by May 1, 1981—or GM would reconsider its options and probably go elsewhere. The city of Detroit would have to pay for the seized property; it would have to pay the relocation costs for more than 4,200 residents who had been evicted; it would have to level all the buildings, clean the entire area of debris, and prepare the grounds for the new plant construction (Wylie, 1989, p. 52). All this meant that it was unlikely that the city would ever recoup its investment in the GM deal, and it had no guarantee that it would ever receive any revenues from GM to invest in schools, garbage collection, police, or other important municipal services (Wylie, 1989, p. 109). However, city officials were between a rock and a hard place. Either they could give in to GM's demands, or they could be branded as antibusiness and lose the GM deal (or so they were convinced by GM representatives).

Everywhere GM goes, the pattern is the same (Dandaneau, 1996). It tells city officials that it will come to their region if the deal is right, forces them to act in haste, continually reminds them that it can go elsewhere if it wants, and then stands back and watches the fur fly. GM executives always express great regret over what is being done in the name of GM, insist that they have great concern for the social ruin that is occurring, continually praise the efforts of public officials, and disavow responsibility for anything that happens. In the Poletown case, the community was characterized by GM executives as a depressed area, filled with dilapidated homes and deteriorating businesses. Executives of GM publicly blamed city officials for the seizure and destruction of the community that would serve as the future site of their Cadillac plant, while they sullied the community and its residents.

General Motors, the city, and other power brokers lined up against Poletown. The powers that be used every device at their disposal to get exactly what they all wanted. Any resistance among Poletowners was branded as unreasonable, immoderate, and vicious, and anything that GM and Detroit officials did was portrayed as normal, legitimate, and in the public interest. A strong ally of the powerful interests that had gathered against Poletown—as is often the case—was the legal institution. A new law had been created in Detroit that made it perfectly legal to take private property for public use if it could be reasonably shown that some public purpose was achieved, and the new law was used for the first time in the Poletown relocation. The law was broad enough that the GM Cadillac plant could be portrayed as something that would serve a public purpose, so Poletowners could be legally evicted. The law also included a "quick-take" clause that made it possible for city officials to seize private property rapidly, relocate the residents, and make the area fit for new construction. Even if former residents had protested and won in court, all they would have received would have been extra monies for their property. They would never get their property returned to them (Wylie, 1989, pp. 54-55). On March 31, 1981, the Michigan Supreme Court ruled in a five-to-two decision that the city acted legally in taking Poletowners' property from them.

The residents of Poletown were not expected by the powers that be to be able to muster much of a defense. Most of them were elderly and working class, and they didn't know what it takes to battle effectively the combined forces of the city of Detroit and one of the most powerful companies on earth. Who does? Despite the size and power of the opponent, Poletown residents did have some things working for them: grit, determination, and a belief in the righteousness of their cause. They believed that America was a place where people who fought against powerful interests had to win in order to prove that the people still had a voice in public affairs.

The saga of the residents' struggle against such an awesome set of opponents is a complex one. They did practically everything that they could to save their community. They wrote letters, held meetings, contacted people who they thought might serve as allies, gave interviews, and publicly protested. Ralph Nader sent five of his workers to Poletown, and some of them became passionately involved in the residents' struggles. Maggie Kuhn, organizer of the Grey Panthers, offered the symbolic support of her organization to the Poletown struggle. Max Gail, star of the once-popular *Barney Miller* television show, spent many days in Poletown, discussing the proposed plant with media representatives and Poletown residents. Some journalists wrote in support of the Poletowners. However, the Poletowners received no major institutional support, so their struggles were probably doomed from the start:

> Residents attempted to contact the chief executives at General Motors, the mayor and his staff . . . , the cardinal and his adjuncts at the archdiocese, the local and high court judges, news editors and station managers, officials at the United Auto Workers, Detroit area clergy, community groups, and members of the Left. None took the time to respond in a genuine manner. None even advocated a meeting of all parties to discuss possible compromises. For a variety of reasons, all of which could be reduced to a collusion of class, the power brokers in Detroit embraced GM's Cadillac project. (Wylie, 1989, p. 84)

The Catholic archdiocese even sold two of the community churches, an act that Poletowners defined as betrayal and that disheartened them mightily.

Residents who had lived in the community a short time, renters, and more youthful residents were the most likely to accept the city's offer, and they left early. Some, however, refused to move under any circumstances. Vandalism and arson increased, and it seemed as if some organized effort was underway to destroy the community so that it could be condemned. The smoke, the fire, and the fear of victimization did eventually drive out practically all of the residents. However, those that remained were a tenacious and determined bunch. They decided to try to save the Immaculate Conception Church from the wrecking ball. Residents moved into the church illegally and barricaded themselves. The city responded by slowing down and eventually stopping services. Phone services were cut off along with water and electricity. The protestors used flashlights and candles and slept on mattresses in the basement.

On July 14, 1981, city officials were fed up, and they moved on the church. Police started setting up metal barricades around it early in the morning. The Special Weapons Attack Team (SWAT) was summoned, and police dogs were used to patrol the area. The police attached a chain to the sanctuary side door and used a tow truck to rip it off its hinges. The protestors fled to the basement and locked the door. The basement door was eventually ripped off, and 20 police officers swarmed into the room, prepared to deal with the protestors. During the eviction, the women—some of whom were in their 70s—were told to leave, and some of the men were arrested, physically attacked, or verbally harassed. When the women refused to leave the basement they too were arrested. The police transported 12 people to the seventh precinct, locked in the back of a paddy wagon. The men were fingerprinted, booked, and eventually moved to a jail downtown, but the women refused to leave the station house. After about 6 hours, both the women and men were released, and all charges were dropped. A small victory, but a victory nonetheless.

With the protestors removed, the demolition of the church could proceed with a vengeance. An 8-foot fence was erected around the church to keep out resisters, and police were dispatched around the perimeter; police helicopters scouted for troublemakers from the air. A crowd of 200 people eventually gathered outside the fence. As the razing of the church proceeded, they prayed, cried, verbally protested, and commiserated with one another. Skirmishes erupted. One man jumped on top of a bulldozer, and he was taken to jail. Some people climbed onto a truck that was trying to remove the church bells, and they were threatened with arrest. Flowers were woven through the fence as a symbolic protest. People pulled on the fence, and police pulled them off and threw some to the ground. A man managed to break into the church, and demolition stopped until he could be found, arrested, and taken to jail. An East Indian family—husband, wife, and two young children—snuck under the fence in the predawn hours, but as they tried to reach the church, they were intercepted by police and arrested. The wife's sari was torn and her glasses were trampled; she was cursed by police, handcuffed, and put in a squad car; and she was told that because of what she had done, her children would be taken permanently away from her. Her husband was hand-cuffed, placed in another police vehicle, and taken to jail. These acts of courage and resistance, however, were little more than dust in the wind. The church was eventually leveled to the ground.

Although the leveling of the church was a tremendous blow to the Poletown-ers' resistance, some of them still maintained the fight. An old GM car was taken to the company's world headquarters and destroyed. Messages had been painted on the vehicle such as "GM Destroys Churches" and "Boycott GM." The protestors took turns kicking and beating the automobile (with feet, hands, or a crowbar), breaking its windows, puncturing its tires, and enacting as much mayhem on the vehicle as they could. Plans had been formulated to burn the vehicle, but in the end it was just abandoned in the street. The pluck of the people who resisted and protested at every turn, whatever else it accomplished, did make visible the close ties and shared interests between the most powerful sectors of U.S. society.

The ultimate irony, and thus the ultimate tragedy, is that none of what happened had to have happened. The plant could have been built, and the community could have been preserved. In fact, some of the Poletown residents who came to be the strongest resisters were not initially against the plant. Why would they be? If the plant brought jobs and led to a revitalization of their community, it would be a blessing to them. Richard Ridley, a Washington architect, devised several plans that would have allowed the plant to exist with the community still intact. Because Poletown was to be destroyed to create either greenspace around the plant or parking areas, several options presented them-selves. What could have been done without too much trouble—*if* the powers that be had been at all willing to bend even a little—was to move the proposed site a bit to north, to rotate it so that it faced in a different direction than the one initially proposed, or to build a multilevel parking structure and reduce or eliminate entirely the amount of greenspace (Wylie, 1989, p. 141). But neither General Motors nor the city council was in an accommodating mood. The protestors called into question the freedom of General Motors to go where it wanted and to do what it wanted, unfettered by any outside restrictions. This "good corporate citizen" had no intention of changing its game plan when it had the resources, interest, and enough influential allies on the team to play real hardball.

White-Collar Rule Breaking: Corporate Skullduggery and Political Racketeering

The term *white-collar crime* was coined by Edwin Hardin Sutherland (1883-1950), and his exploration of this form of deviance is the distinguishing mark of a distinguished career. The American Sociological Society (ASS) held its 34th annual meeting jointly with the American Economic Society's (AES) 52nd meeting, in December 1939 in Philadelphia. There the president of the AES, Jacob Viner, addressed his audience about the relationship between public policy and economic doctrine (Viner, 1940). After he finished, Sutherland (1940), the new president of the ASS, advanced to the podium and spoke about white-collar crime. (The group wisely changed its name to the American Sociological *Association* once it became fashionable to use acronyms for professional organizations.)

Sutherland defined white-collar crimes as crimes committed by persons of respectability and high social status in the context of their professional activities, occupations, or jobs. He included crooked practices in occupations, professions, businesses, corporations, and even politics in this category of crimes. He traced the causes of white-collar crime to the structure of a society. Differential social organization and differential association create a situation in which some people encounter group supports and the motives, rationalizations, techniques, and definitions that make crime more likely. Sutherland believed that the white-collar offender holds two positions that are in conflict. The demands of a business organization can create in-group loyalties that can easily lead to an indifference to the demands of legal or moral norms and to an insensitivity to the interests or needs of other groups such as the public, customers, or even the employees of the business organization.

The intended message of Sutherland's address was clear to his audience: White-collar crime is real crime, it causes tremendous destruction, the perpetrators have a wanton disregard for the consequences of their actions, and the victims of it are both unsuspecting and severely damaged. White-collar crime rips the fabric of society, Sutherland believed, and its perpetrators deserve the ultimate in scorn, derision, and punishment. More than scientific interest prompted his interest in white-collar crime. He was outraged by the egregious acts of corporate representatives, and his sense of social decency was offended (Green, 1997, p. 6).

Sutherland's own experiences in getting his ideas into book form seem to support much of what he believed about the benefits of privilege and power (Geis & Goff, 1983). He had amassed a great deal of information about the collective and individual harm caused by some of the largest and most powerful companies on earth. His writings about the predations of these organizations were filled with the names and histories of the offending companies. Editors at Dryden Press balked at Sutherland's intention to describe the companies as criminal even though none of them and none of their representatives had ever been actually convicted of any crimes. The administrators of the university where Sutherland taught, Indiana University, in Bloomington, Indiana, were also made skittish by Sutherland's plan to reveal in writing the identities of offending corporations. They apparently feared that the donations from these companies might stop. Sutherland eventually succumbed to the pressure and allowed his work to be published without the names of the offending companies (Sutherland, 1949). It was not until 1983 that a different publisher offered the uncut version of Sutherland's book in which the names were finally revealed (Sutherland, 1983).

Many years after Sutherland's address to the ASS on white-collar crime, Clinard and Quinney (1973, p. 188) discerned two distinct types of crime covered by Sutherland's omnibus term: corporate and occupational. The crux of the distinction is who benefits from the antisocial activity. Corporate criminal behavior directly helps an organization itself, even though agents of the corporation carry out the acts. Occupational criminal behavior is done by a person of respectability and high social status who is him- or herself the primary beneficiary. Even if the rule breaking takes place in a group setting, the deviant is acting primarily on his or her own to further his or her own selfish interests, and the organization is not responsible; in fact, the organization itself can be a victim (Belbot, 1995).

Contemporary writers insist that white-collar crime should be defined in terms of the respectability and legitimacy of the position rather than in terms of the color of an employee's collar. Because many of the most egregious acts are actually carried out by people who are not particularly powerful or wealthy but who do have respectable positions in business or government, it may be better to think in terms of positions rather than people. In addition, because white-collar crime is not the prerogative of the corporate sector—it can be found in politics, the military, labor unions, medicine, law, universities—it may be wise to replace the words *corporate crime* with the words *organizational crime* (Coleman, 1998, p. 12). People who commit white-collar crimes may have certain personal characteristics that separate them from some of their more law-abiding colleagues: low self-control, a search for immediate gratification, and an indifference to the long-term consequences of their actions (Gottfredson & Hirschi, 1990, pp. 190-191).

In most of the white-collar crimes that have plagued us through the ages, the distinction between occupational and organizational crime is irrelevant because *both* the organization and its representatives benefit from the ongoing deviances. These normative violations help the organization and its members to compete in a system where survival is never guaranteed and where objective indicators of personal and professional success are few and far between. Large corporations have become both the sites of acts of major rule breaking and the "instruments" used to carry them out (Tillman & Pontell, 1995). Some individuals are able to use the preeminent legal status and financial resources of the corporate structure to accomplish their own objectives and to impose their will on others (Roy, 1997, pp. 267-268).

The text of Sutherland's address to the ASS shows that he believed that politicians presented a lesser threat than representatives of the business world (Sutherland, 1940, p. 4). I wonder if he would still think this way if he were alive today. It is hard to be at all familiar with political rackets and not have a suspicion that far too many politicians view the government as their own private cash cow that can be milked whenever the need arises. If there is a difference between political rackets and other kinds of trespass, it is to be found in opportunity structures, levels of accountability, and organizational supports, not in the greater morality or integrity of politicians as compared to other kinds of white-collar crooks.

Martin Gross (1996) gave us a glimpse of the underbelly of political life in his exploration of corruption, duplicity, and self-interest in American politics. He showed how the arrogance of power and a shield of elitist invisibility can work together to allow destructive acts to flourish unchecked and undefined because of the power of who does them and the sophistication with which they are committed. The great irony in the study of deviance—which, however, is of no surprise to relativists—is that certain people can cause great harm without being defined or having their actions defined as harmful. The structure of relations in political organizations blurs responsibility, and it is difficult for an outsider to know who did what. Top officials deny responsibility by claiming that they had no knowledge of how things were being implemented, and low-level employees deny responsibility by claiming that they were following orders and remaining loyal to their superiors. Regardless of outcome, blame is placed on individual actors and not on organizational structure (Cavender, Jurik, & Cohen, 1993, pp. 157-159).

Politicians have been pilloried for campaign corruption, lying, indifference to the needs of voters, conceit, cronyism, ignorance, and mismanagement of the governmental apparatus. Their greatest sins may be their incredible waste of valuable resources and their taking and squandering (under the sanction of law) of other people's belongings (Gross, 1996, pp. 7-8). Almost every politician tries to reward special constituents by throwing special projects their way ("pork"). This practice is rampant, almost wholly unaccountable, and an example of stealing without stigma. Consider some examples of "pork" (Examples 1–5 from Gross, 1996, pp. 133-134; Example 6 from Thomas, 1996):

1. $2 million was sent to Toledo, Ohio, to assist the Farmer's Market.

2. $125,000 was given to support a summer program for school teachers on how photography influences the American identity.

3. $161,913 was allotted to study how the Israelis responded to Iraqi SCUD attacks. (They weren't popular.)

4. $105,163 was allocated to study monogamy among biparental rodents.

5. $5.1 million was apportioned to build a third golf course at Andrews Air Force Base near Washington.

6. The U.S. Senate kicked in $1 million to make it possible for scientists to study the brown tree snake, a reptile found only in Guam, harmless to humans, and incapable of surviving anywhere in North America.

Politicians take our money from us, money we may very well not wish to relinquish, and they then use it in ways over which we have no control. How is this any different from the depredations of a garden-variety thief? If a difference can be found between pork and theft, it is found primarily in who has the power to say what is legal and what is not.

Sutherland (1940) claimed in his presidential address that the "inventive geniuses" for many kinds of white-collar crime are attorneys (p. 11). If only he could see us now, he would realize how prophetic his words have turned out to be. Without the assistance of powerful and influential lawyers who can manipulate and often subvert legal procedures and offer advice, encouragement, and sometimes direct support to those privileged and powerful enough to be able to afford their services, it is unlikely that the wealthy would fare as well as they have (Lauderdale & Cruit, 1993). Attorneys make it possible for some people to take other people's belongings with little or no accountability (Nader & Smith, 1996).

Members of the upper class, Sutherland insisted, have been successful in keeping many of their jeopardizing or harmful acts out of criminal court. This may be the principal difference between the trespasses of the upper class and the trespasses of everybody else.

> The crimes of the upper class either result in no official action at all, or result in suits for damages in civil courts, or are handled by inspectors, and by administrative boards or commissions, with penal sanctions in the form of warnings, orders to cease and desist, occasionally the loss of a license, and only in extreme cases by fines or prison sentences. (Sutherland, 1940, pp. 7-8)

As a result of this special treatment, Sutherland declared, white-collar crooks are not viewed as real criminals by the general public or even by themselves. They may have done something technically illegal, but they do not believe that they are real criminals.

Having their misdeeds regularly omitted from the dockets of the criminal court apparently is not enough for the privileged and powerful. They have worked to ruin the civil justice system as well. These are nothing less than audacious efforts by corporate representatives (and politicians who represent their interests) to keep themselves free from harm, despite the great injury that their profiteering has done to the individuals who have been adversely affected by environmental pollution, defective automobiles, unsafe appliances, carcinogens in tobacco, and other dangerous products (Nader & Smith, 1996, p. 259). Money, hype, and monotonous repetition are used to perpetuate a false image that corporations get railroaded into civil courts where they face a barrage of frivolous lawsuits and pay huge and undeserved fines. The truth is that corporations are *not* the target of a flood of lawsuits from consumers. If an increase in lawsuits against businesses does exist, it is due to businesses suing other businesses, a form of litigation that will undoubtedly continue (Nader & Smith, 1996, pp. 262-314). Juries are far from antibusiness, and they award punitive damages to an injured citizen only if it is clear that corporate representatives acted with glaring disregard for the consequences of their actions (Nader & Smith, 1996, pp. 278-279).

Stella Liebeck, age 79, was a front-seat passenger in her grandson's car. They pulled into a McDonald's restaurant in Albuquerque, New Mexico, so that Ms. Liebeck could order coffee. She tried to remove the lid from the cup to add cream and sugar—the vehicle was motionless—and the entire contents spilled on her lap. She experienced third-degree burns on her thighs, groin, and buttocks, and she was permanently scarred on 16% of her body. These injuries were serious, and it took more than 2 years for her to recover fully. Ms. Liebeck wanted some compensation for her emotional ordeal, some help with her medical bills, and some recognition by McDonald's of the dangers lurking in those cups of coffee for other unsuspecting customers. However, when she appealed to restaurant executives for help, they were uncompromising. They refused to pay anything close to what her operations had cost ($15,000 to $20,000); all they did was to offer Ms. Liebeck $800. In desperation, she hired a lawyer, and a lawsuit was filed.

At the trial, at least initially, Liebeck faced a skeptical jury. Members of the jury panel believed that a lawsuit over a cup of coffee was a waste of their time, frivolous, and malicious. They certainly had no discernible animosity toward McDonald's. However, as the details of the case unfolded in the week-long trial, McDonald's reputation was more and more tarnished. Testimony was introduced that corporate headquarters required that the temperature of the coffee be maintained at between 180 and 190 degrees Fahrenheit to make it taste better for longer periods. (Many other restaurants keep the temperature of their coffee at 160 degrees.) At this high temperature, severe burning and tissue destruction occur within seconds in the case of a spill. Coupled with an unreasonably high temperature was the quality of the McDonald's cup: It has exceptional insulating properties, so coffee drinkers are unable to appreciate just how hot the coffee really is. The jury concluded that McDonald's did keep the temperature of the coffee much too high for human consumption and that customers were given insufficient warnings about the potential dangers. The jury settled on compensatory damages of $200,000, which it then reduced to $160,000 because it believed Ms. Liebeck had to bear some of the responsibility for spilling the coffee on herself.

The jury awarded an additional $2.7 million in punitive damages to Ms. Liebeck. Why? The answer is simple: McDonald's had shown a wanton disregard for the well-being of its customers. It seems that 700 other people had experienced a fate similar to Ms. Liebeck's between 1982 and 1992; even some children had been severely burned in accidents involving the too-hot coffee when cups were knocked over or dropped and the contents spilled on the youngsters. Executives of the restaurant chain believed that no compelling reason existed to turn down the heat because, considering all the coffee that their restaurant chain served, 700 burns were insignificant. The jury concluded that anyone who ordered a cup of the too-hot coffee was a potential burn victim. The jury's principal objective was to get McDonald's attention and force its executives to do the responsible thing: reduce the temperature of the coffee. The jury came up with the $2.7 million figure because that is how much McDonald's makes from 2 days of selling coffee.

Although the size of the punitive damages does not seem at all excessive considering the size of the McDonald's organization and earning power, the judge decided that the punitive damages were too high. He reduced the punitive damages to $480,000, three times greater than the compensatory damages. McDonald's had intended to appeal the $640,000 award, but it finally agreed to settle for an undisclosed amount if the details of the arrangement were kept silent forever by Ms. Liebeck and her attorney. The McDonald's restaurants in Albuquerque are now serving the coffee at 158 degrees, so maybe Ms. Liebeck's efforts did achieve the desired result. However, the award that she received will hardly be sufficient to compensate her for all that she endured. (The account of this case is based on Nader & Smith, 1996, pp. 266-273.)

Flexible Penalties

Sutherland's (1949) analysis of white-collar crime showed him that its perpetrators were able to avoid the stigma and the punishment accorded lesser offenders. The legal status of white-collar crimes is more ambiguous, and laws may be more difficult to apply to incidents of white-collar wrongdoing (Gerber & Weeks, 1992, p. 330). Obviously, behavior is illegal only if laws exist to criminalize it. Laws are crafted by lawmakers, and lawmakers can be influenced and their outlooks refurbished by powerful people who work to control the content of law and how it applies to them (Livingston, 1996, pp. 294-300).

The arrogance of power can easily prompt an individual to believe that the rules and laws of ordinary folk are inapplicable to him or her. This heightened sense of importance and invincibility can motivate and rationalize many untoward activities. Richard Nixon, forced from the office of president of the United States for his role in the Watergate scandal, claimed—and probably believed—that if the president did it, it was not against the law. Key players in Contragate, in which arms were sold to Iran—illegally, some people claimed—and some of the monies were used to aid the Contras in Nicaragua, assumed that everything that they did was legal.

> Organizations have their own versions of events and respond to accusations of deviance with accusations and assertions of their own. In the exchange of claims and counterclaims, whether an organizational action actually comes to be seen as deviant is fundamentally a matter of definition. (Ermann & Lundman, 1996, p. 25)

The vacuum created by disputes over the legal status of some untoward act has great benefits for white-collar offenders who capitalize on the ambiguity and turn it to their own ends.

Even when laws are firmly in place, it may be more difficult to detect white-collar crimes, arrest and convict white-collar criminals, and then punish the organizations for which they work. Because victims may be unaware that they have been intentionally hurt by some misdeed, or because an offending corporation has neither a body to kick nor a soul to damn, it is more difficult to apply to white-collar misdeeds the same legal procedures and penalties that were invented to cover one-on-one harms of individuals (Belbot, 1995, pp. 229-232). The nature of law, the size and power of the corporation, the privileged status of white-collar offenders, and the diffuse nature of white-collar victimizations all mean that penalties for white-collar crime are likely to be relatively painless and ineffective. A fine of $3 million to a company such as McDonald's, even if it had been paid, would have been little more than a slap on the wrist.

What would be an effective and just penalty for taking other people's belongings? Perhaps Islamic law can serve as an inspiration. It requires that a thief will have his or her hand amputated to serve both as a just punishment and as a deterrent to the thief and to others who might be tempted to transgress against the community. A hand amputation is done in a public and central place, usually the center of town, so that people will have the opportunity to witness the event. The offender is led to the designated spot, the verdict is read aloud, and the arm is stretched out on the surface of a table. A person skilled in amputation grabs the hand, exerts pressure on it with a sharp knife, and severs it from the body. A male physician and a male nurse then quickly stop the bleeding by applying a bandage to the wrist of the amputee. No matter how many people witness an amputation, they always maintain a somber and dignified mood throughout the event (Souryal, Alobied, & Potts, 1996, pp. 436-437).

The basis of social order in Islam society is a set of norms that sanctify social decency and emphasize the importance and preeminence of the community. Theft is viewed as a serious offense, so severe punishment is viewed as necessary and beneficial for all concerned. Theft destroys communal living by threatening personal property and keeping citizens in fear for their freedom to own or to trade with others (Souryal et al., 1996, p. 433). The swift and certain punishment of hand amputation is justified not only as a just penalty for a particular offense but also as a more general deterrent to the commission of more serious crimes of murder, rape, or robbery that might flourish if lesser crimes were not nipped in the bud. Hand amputation teaches the seriousness of betraying others' trust and of violating community standards; it corrects an individual who is indifferent to the damage he or she has done; and it gives an individual who has done wrong quick penance and almost immediate reintegration back into the community. After the amputation, the individual is freed and considered to have paid his or her debt to the community (Souryal et al., 1996, p. 440).

Summary and Conclusions

Taking other people's belongings exists in a social context. This taking is powerfully influenced by power and privilege, as is the meaning of this taking. Whether it be a simple theft such as three-card monte or the more complicated and injurious depredations of white-collar deviants, almost always a web of relationships makes the deviance what it is. Power is a critical variable in understanding this kind of deviance. What is defined as criminal or deviant and what penalties are recommended reflect the power and interests of specific groups. If you are a carrier, your taking will be dealt with harshly; if you are General Motors, your taking may not be viewed as taking at all, at least by the groups that benefit from the taking. Some people (or the organizations that they represent) can take with impunity, whereas others will have a hand amputated for their efforts. Theft is relative, and penalties are flexible. Some people can take things of great value from others. Not only may it be legal, but it may be viewed as a valuable renovation project.

References

Belbot, B. (1995). Corporate criminal liability. In M. Blankenship (Ed.), *Understanding corporate criminality* (pp. 211-237). New York: Garland.

Box, S. (1983). *Power, crime, and mystification.* London: Tavistock.

Cavender, G., Jurik, N., & Cohen, A. (1993). The baffling case of the smoking gun: The social ecology of political accounts in the Iran-Contra affair. *Social Problems, 40,* 152-166.

Chambliss, W., & Seidman, R. (1982). *Law, order, and power* (2nd ed.). Reading, MA: Addison-Wesley.

Clinard, M. B., & Quinney, R. (1973). *Criminal behavior systems: A typology* (2nd ed.). New York: Holt, Rinehart & Winston.

Coleman, J. W. (1987). Toward an integrated theory of white collar crime. *American Journal of Sociology, 93,* 406-439.

Coleman, J. W. (1998). *The criminal elite: Understanding white-collar crime* (4th ed.). New York: St. Martin's.

Dandaneau, S. (1996). *A town abandoned: Flint, Michigan, confronts industrialization.* Albany: State University of New York Press.

Davis, N. (1980). *Sociological constructions of deviance: Perspectives and issues in the field* (2nd ed.). Dubuque, IA: W. C. Brown.

Ermann, M. D., & Lundman, R. (1996). Corporate and governmental deviance: Origins, patterns, and reactions. In M. D. Ermann & R. Lundman (Eds.), *Corporate and governmental deviance: Problems of organizational behavior in contemporary society* (5th ed., pp. 3-44). New York: Oxford University Press.

Geis, G., & Goff, C. (1983). Introduction. In E. H. Sutherland, *White collar crime: The uncut version* (pp. ix-xxxiii). New Haven, CT: Yale University Press.

Gerber, J., & Weeks, S. (1992). Women as victims of corporate crime: A call for research on a neglected topic. *Deviant Behavior, 13,* 325-347.

Gottfredson, M., & Hirschi, T. (1990). *A general theory of crime.* Stanford, CA: Stanford University Press.

Green, G. (1997). *Occupational crime* (2nd ed.). Chicago: Nelson-Hall.

Gross, M. (1996). *The political racket: Deceit, self-interest and corruption in American politics.* New York: Ballantine.

Hall, J. (1952). *Theft, law, and society* (2nd ed.). Indianapolis: Bobbs-Merrill.

Lauderdale, P., & Cruit, M. (1993). *The struggle for control: A study of law, disputes, and deviance.* Albany: State University of New York Press.

Livingston, J. (1996). *Crime and criminology* (2nd ed.). Upper Saddle River, NJ: Prentice Hall.

Maurer, D. (1974). *The American confidence man.* Springfield, IL: Charles C Thomas.

Nader, R., & Smith, W. (1996). *No contest: Corporate lawyers and the perversion of justice in America.* New York: Random House.

Poveda, T. (1994). *Rethinking white-collar crime.* New York: Praeger.

Prus, R. (1977). *Road hustler: The career contingencies of professional card and dice hustlers.* Lexington, MA: Lexington.

Quinney, R. (1970). *The social reality of crime.* Boston: Little, Brown.

Roy, W. (1997). *Socializing capital: The rise of the large industrial corporation in America.* Princeton, NJ: Princeton University Press.

Scarne, J. (1974). *Scarne's new complete guide to gambling* (Rev. ed.). New York: Simon & Schuster.

Simon, D. (1996). *Elite deviance* (5th ed.). Boston: Allyn & Bacon.

Souryal, S., Alobied, A., & Potts, D. (1996). The penalty of hand amputation for theft in Islamic justice. In C. Fields & R. Moore, Jr. (Eds.), *Comparative criminal justice: Traditional and nontraditional systems of law and control* (pp. 429-452). Prospect Heights, IL: Waveland.

Sutherland, E. H. (1940). White-collar criminality. *American Sociological Review, 5,* 1-12.

Sutherland, E. H. (1949). *White collar crime.* New York: Dryden.

Sutherland, E. H. (1983). *White collar crime: The uncut version.* New Haven, CT: Yale University Press.

Thio, A. (1995). *Deviant behavior* (4th ed.). New York: HarperCollins.

Thomas, C. (1996, April 1). Congress hasn't cut pork from the menu. *Lexington Herald-Leader* (Lexington, KY), p. A7.

Tillman, R., & Pontell, H. (1995). Organizations and fraud in the savings and loan industry. *Social Forces, 73,* 1439-1463.

Viner, J. (1940). The short view and the long in economic policy. *American Economic Review, 30,* 1-15.

Wylie, J. (1989). *Poletown: Community betrayed.* Urbana: University of Illinois Press.

8

Drugs and Drug Taking

Introduction: The Social Reality of Drug Use

Drug taking is a complex process, and understanding it requires a knowledge of more than drugs. Every drug experience reflects a user's expectations and experiences with regard to a chemical substance, as well as the setting—where a chemical substance is used, with whom, when, why—and the entire socio-cultural environment and its repository of meanings and understandings (Weil, 1972). All these factors introduce a great deal of variability into any drug experience.

> To assume that the biochemical properties of a drug are the sole or major cause of drug behavior ignores the cultural and social variability of drug behavior and experience. In order to understand this variability, we must examine the individual's expectation of drug effects and the physical and social environment in which the drug is taken. (Matveychuk, 1986, p. 7)

Molecular structure of a drug is one factor that will help to determine the effects a drug will have on mind and body, but pharmacological effects of any drug are always mediated by social meanings, social contexts, and social relationships. The reality of drugs, drug use, and drug abuse is a socially constructed one.

Imaging Drugs and Drug Abuse

What is a drug? Get any textbook on drugs, look up the term, and you will find that it is defined as a chemical substance that when ingested alters or changes the functioning of the mind or body (Levinthal, 1996, p. 4). A textbook definition of a drug gives the impression—incorrectly, as it turns out—that drugs have uniform and universal characteristics that make it possible for them to be classified as drugs. The truth is that no single uniform feature is found in all the substances called drugs that differentiates them from all the substances called nondrugs (except that all drugs have been called drugs by somebody).

> Some drugs, such as heroin, alcohol, valium, nicotine, and caffeine, may lead to withdrawal symptoms upon discontinuance of use; but drugs such as peyote, LSD, and marijuana do not. Some drugs, such as cocaine and amphetamines, stimulate the central nervous system while others, such as barbiturates and alcohol, depress it. Some drugs have medicinal value, others have none at all. In fact, efforts to argue that drugs are defined by their action on our bodies or mind[s] lead only into absurdity since everything we eat and drink is a chemical that effects our bodies and minds. We can only conclude, then, that the single characteristic that drugs have in common is their classification as drugs. (Matveychuk, 1986, p. 9)

The naming of some substance as a drug is a highly relative, value-laden, often personal judgment, and it cannot be separated either from the social context or from who is doing the defining. Practically any claim about drugs is contentious, and discussions about drugs are almost always highly charged.

The original Greek word for drug, *pharmakon,* contained three images: the substance was a poison (bad), a remedy (good), or a magical amulet (wonderful). Drugs and drug use have been viewed in many different and often contradictory ways: demonic, deadly, enslaving, monstrous, curative, spiritual, liberating, fun, interesting (Montagne, 1988, pp. 418-419). One group's poison is another group's medicine, and sometimes the same substance can be both a poison *and* a medicine to the same group (or to select individuals in that group). Cultural meanings and personal experiences work together to help users and nonusers alike to perceive, interpret, recall, and discuss drug-taking experiences, and they influence the course of an individual's drug-taking career. "These personal, social, and media metaphors, which provide descriptions of drug effects, are very influential in helping the user perceive, interpret, and respond to the pharmacological actions of drugs. They also assist in structuring ways of remembering these experiences" (Montagne, 1988, p. 420). Sometimes these generalized images even account for the actual effects of a drug on a user (Montagne, 1988, p. 420).

If the term *drug* is slippery, then the term *drug abuse* is even more slippery. Pronouncements about the abuse potential of a chemical substance function to reinforce some group's particular conception of reality and to justify the persecution of people who do not share that group's view of things (Szasz, 1985). Even if it is true that the use of a particular chemical substance can be harmful to users—what chemical substance couldn't be abused if the user were to try hard enough?—the potential for harm is not the principal reason that one drug is branded as a substance of abuse and another is not. Scientific evidence about the harmfulness of a particular drug almost always comes *after* the declaration by authorities that it is undeniably a drug of abuse (Szasz, 1985). "People in positions of social responsibility have greater credibility and authority to prove, create and maintain their reality, while some have no credibility and must suffer the consequences of the reality created by others" (Matveychuk, 1986, p. 11). Those societies that have an extensive user culture are the ones most likely to provide users with positive images of the drug experience. Where this user culture is absent or poorly developed, an antidrug message is more likely, and drug effects are more likely to be portrayed as dangerous and debilitating; even users themselves may hold negative views of their own drug experiences (Becker, 1967).

The classification of some chemical substance as a drug of abuse—and its use as deviant—is not determined exclusively by qualities intrinsic to the drug or even its effects. Situational factors seem to be very influential. Respondents in one investigation were asked to decide on the acceptability or unacceptability of the use of alcohol and marijuana in various hypothetical situations. The details of a user's intent, motive, goals, location, and patterns of use were varied by the researcher to see if this would change respondents' attitudes about the deviancy of drug use. The major finding was that situational factors did explain a "considerable degree" of the variance in how respondents viewed the use of alcohol and marijuana (Orcutt, 1996, p. 221). Consider the following:

> The use of antihistamines to avoid cold symptoms seems unlikely to be defined as deviant, that is, unless the user is a Christian Scientist or a member of any one of several other religions that advocate treating disease by spiritual means only. Drinking alcohol is permitted in our society, but its manufacture and sale were once illegal nationwide, and in some areas it is still illegal to sell alcoholic beverages. Everywhere else in the United States the distribution and sale of alcohol are highly regulated by laws specifying where, when, and to whom it is sold. Amphetamines are widely available through medical prescription, but taking them without a prescription makes one liable to imprisonment for using a dangerous drug. (McCaghy & Capron, 1997, pp. 283-284)

The deviancy of drug use is in the eyes of the beholders, and judgments are characterized by a great deal of relativity.

Getting Stoned in the Animal Kingdom

Our human ancestors may have learned much of what they knew about intoxication from observing the kinds of plants that nonhuman animals consumed and their effects. Getting stoned in the animal kingdom probably is influenced by something physiological because certain substances do seem to produce similar effects among dissimilar organisms.

> After sampling the numbing nectar of certain orchids, bees drop to the ground in a temporary stupor, then weave back for more. Birds gorge themselves on inebriating berries, then fly with reckless abandon. Cats eagerly sniff aromatic "pleasure" plants, then play with imaginary objects. Cows that browse special range weeds will twitch, shake, and stumble back to the plants for more. Elephants purposely get drunk on fermented fruits. Snacks on "magic mushrooms" cause monkeys to sit with their heads on their hands in a posture reminiscent of Rodin's *Thinker.* (Siegel, 1989, p. 11)

Siegel (1989) labeled intoxication the fourth drive, little different from other basic drives such as hunger, thirst, or sex (p. 10). He stated that the fourth drive does not simply motivate organisms to consume external substances to feel good. It motivates them to attain some altered state, and the direction of that change—up or down or good or bad—is of secondary importance (p. 217). Sometimes nonhuman animals will stop their regular activities—eating, drinking, sex—and take psychoactive chemicals if they have the chance (p. 64).

Studies of the drug use of nonhuman animals in the wild are interesting. They suggest that a part of the variation in drug-taking experiences is accounted for by the relationship between biological structures and chemical substances. However, naturalistic studies are flawed. They credit nonhuman animals with far too many human qualities, and they suffer from insurmountable problems of interpretation and measurement. It is impossible to know for sure exactly what is going on when an animal gets intoxicated. Because little is known (or knowable) about an animal's needs, its unique biological characteristics, the quality of its group life, how much of a psychoactive drug is in its body, its stresses and strains, its specific body weight, its fat composition, its psychological characteristics, or its temperament, the demonstration that some nonhuman animal is intoxicated has little that it can really tell us about drugs, drug use, or drug abuse.

Laboratory studies of drug-taking behaviors of nonhumans, though more artificial than studies in the wild, are helpful because they allow more precision in determining how drugs work. Test animals in laboratories do engage in drug-taking behaviors when given the opportunity. Monkeys allowed to press a bar to get an injection of morphine (a central nervous system depressant) will press it hundreds of times for one injection even if they only receive a small amount of the drug (Ray & Ksir, 1996, pp. 41-42). Rats, dogs, baboons, and several species of monkeys will go through a series of complicated physical maneuvers and forego practically all other pleasures if cocaine is the reward. A laboratory animal might be forced to press a bar 100 times for its first injection. On subsequent trials, the number of presses is then doubled to 200, 400, 800, and so on. At some point, an animal will either stop responding or be too exhausted to continue, and this is called the breaking point. Cocaine's breaking point is 2 to 16 times higher than that for other drugs, and some monkeys have been found to be willing to press a bar 12,800 times for one cocaine injection (Siegel, 1989, pp. 183-185).

Lab studies clearly show that certain drugs are powerfully reinforcing, even irresistible, to laboratory animals with free access to them. When some lab animals are given the opportunity for unlimited access to some chemical substances, they indulge themselves to the point where other pursuits become irrelevant (Ray & Ksir, 1996, p. 42). What lab studies mean, however, is not as clear. These studies do not prove that some inevitable, inexorable, physical addiction process exists to account for compulsive drug use among members of the animal kingdom (Peele, 1985, p. 77). The lab animals may self-administer drugs simply because they are attached to implanted catheters that make it easy for them to get drugs and are frightened, under stress, in pain, and living in overcrowded, uninteresting, and unfamiliar environments (Peele, 1985, p. 79). Because these studies maximize the chance that a hapless creature will self-inject drugs, they really do not prove that drug-taking behavior is rigidly determined by the properties of drugs (Peele, 1985, p. 94). If animal research has any relevance for understanding human drug experiences, it is in its demonstration that one important source of a drug habit is a life filled with wretchedness. Humans may develop a drug habit when they do not have the kinds of rewarding experiences that make a drug-free life worth living, either because of their own lack of effort or because of circumstances over which they have little control (Peele, 1985, p. 96).

Arbitrary Addictions

How we define or label a human activity affects greatly the way we react to it. If drug use is classified as an addiction and the user as an addict, this designation alone makes it much less likely that the drug use will be viewed as anything other than a sickness or disease. If it were to be defined in some other way—sin, psychological malfunction, a reaction to social disorganization, a problem in living, a central activity, a form of self-help, a variation on ordinary behavior, or ordinary behavior itself—it would generate entirely different understandings and evoke entirely different reactions. Representing drug use as an addiction makes it appear inexorably and inherently compulsive, self-destructive, and in need of cure.

The labels "drug addiction" and "drug addict" are much more than an objective classification of certain kinds of activities and people. Before the 20th century, the word *addiction* was used to refer to a habit that could be *either* good or bad but was most often good. Back then, addiction meant no more than a preference, interest, or inclination that people had. They might be addicted to religion, hard work, reading, sobriety, alcohol, money, their families, or even the devil (Szasz, 1985, pp. 6-7). Even though people did not always approve of one another's so-called addictions, the word had not yet attained its meaning as a relentless and self-defeating pursuit. The words "addict" and "addiction" eventually underwent a transformation and were expanded to cover a large number of people and activities that fit only loosely, if at all, the original understanding of "addict" and "addiction."

Changes in the meaning of "addiction" and "addict" always produce changes in the kinds of substances that are defined as addicting, the kinds of drug-taking patterns that are viewed as addicting, and the kinds of people who are classified as addicts. Some drugs that were once considered innocuous and benign are now considered dangerous and addicting, and some drugs that were once considered addicting and dangerous no longer are. Controversy and dispute seem to be the rule, and different groups will hold vastly different understandings about drug addictions and drug addicts. "The problem is that, although some drugs may be more likely than others to result in addiction, there is no clear distinction between substances that are addicting and those that are not" (Ray & Ksir, 1996, p. 43). Clearly, more is involved in the determination that an individual is addicted to an addicting drug than close scrutiny of a drug's chemical properties (Staley, 1992, p. 99).

The meaning of addiction continued to change throughout the 20th century until it reached a point where it meant not only something generally bad but something invariant, uniform, and universal (Acker, 1991). Though the words "addict" and "addiction" are judgmental and arbitrary, some people have labored to make these terms seem more objective and factual by identifying physical changes—tolerance and withdrawal—supposedly associated with drug use. Drugs such as alcohol and heroin that were correlated with definite physical changes were branded as addictive, and other drugs such as marijuana and cocaine were classified as nonaddictive and only habituating because users experienced neither tolerance nor withdrawal symptoms.

Drug use patterns changed in the 1960s and 1970s, and the meaning of "addiction" changed right along with them. Drug patterns were becoming more diverse, and more and more people were using more and more drugs. The traditional meanings of "addict" and "addiction" were considered too restrictive by authorities to include all the new drug users and patterns of drug use that they believed needed to be included. Addiction was transformed into more of a psychological process than it had been, but a psychological process that still had universal biological underpinnings. By the 1980s, drugs such as cocaine and marijuana were included in the category of addicting substances. A person was defined as addicted if he or she craved a drug, used it uncontrollably, and remained indifferent to the actual or potential adverse consequences from its use. It became the prevailing scientific view that psychological dependence, based on reinforcement, was the driving force behind all drug addiction; physical changes could still occur, but these were no longer considered necessary to identify an addiction or an addict (Ray & Ksir, 1996, p. 42).

Not only is the new definition of "addiction" boundless—applicable to practically anything that anybody wants somebody else to change—but it is more biomedical than ever. The reinforcement that addicts receive, the story now goes, comes from a jolt of pleasure and exhilaration associated with the flow of brain chemicals (Volkow, Wang, Fischman, et al., 1997). Experiences of the addict such as craving, compulsion, and loss of control may also be due to these same brain chemicals (Volkow, Wang, Fowler, et al., 1997). The prevailing view (as of this writing) is that practically anyone can become addicted to practically anything under the right circumstances because of something going on in one's brain.

We can appreciate some of the difficulties with the terms "addiction" and "addict" by looking at their application to a type of behavior that at one time would never have been called an addiction: human sexuality. Are some people addicted to sex, or is the concept of sex addiction nothing more than a silly fad? Opinions vary, and different groups have very different answers to these questions.

Some people believe that no such thing as a sex addiction exists. Sexual behavior is not a substance, they claim, and it is not associated with tolerance and withdrawal symptoms at too-low doses. To call erotic feelings and behaviors addictions is to brand them as abnormal and as not only uncontrolled but

uncontrollable (Henkin, 1996). Does this contribute anything to our understanding of either addiction or sexuality, or does it merely stigmatize those people
who are unhappy with their sexuality but feel powerless to change it?

Other people, however, believe wholeheartedly in the existence of sex addictions. According to them, a sex addiction is a life-threatening condition where
people's lives are dominated by a pattern of compulsive, destructive sex that they
would like to change but cannot. Sex can make you high, they claim, by releasing
chemicals in the brain (Rice, 1998, p. 218). Some people reach the point where
they require, no matter what happens, the rush or excitement that comes from
some forbidden sexuality in order to feel normal and satisfied. Sex addictions
are said to lead to suicide, unwanted pregnancies, family disintegration, violence,
dramatic health care costs, and child abuse. People who refuse to believe that
sexuality can be an addiction are themselves branded as suffering from an illness:
the denial of an illness (Carnes, 1991, pp. 9-38).

A messy term such as "sex addiction" gets even messier when it is applied to
concrete cases. Who is a sex addict? This category seems to include a diverse
group of people whose only qualification is that they have had their lives made
difficult, tragically so in some cases, by sex. Carnes (1991) offered the following
as actual examples of sex addicts and their addictions: a woman who had burned
herself using a vibrator had to go to an emergency room for treatment; a priest
who stole money from his parish to pay for visits to prostitutes; a dentist who
drugged his unresponsive wife in order to have sex with her; a corporate
technician who sexually harassed people; and a youth leader who had sex with
boys (pp. 9-10).

Some people do use the terms "addiction" and "addict" indiscriminately to
condemn whatever they think is bad, compulsive, and in need of correction,
regardless of whether they are discussing sex, drugs, or some other activity that
they do not like. When the brush sweeps this wide, it is probably time to rethink
seriously our understanding and use of the terms "addiction" and "addict." Use
of the addiction terminology does incorrectly make some things look intrinsically good and other things look intrinsically bad. Personal responsibility for
personal behavior such as drug taking or sexuality becomes less credible with
every new discovery of a biomedical cause of addiction, and so-called addictions
are implicitly excused by the tacit understanding that they are uncontrollable
(Szasz, 1985: xii).

Our most damaging addiction may be our addiction to the addiction terminology (Henkin, 1996, p. 64). The availability of the terms "addiction" and "addict"
may offer some advantages to a person who is doing something that seems

compulsive and damaging to others or to self. However, faith in the addiction terminology can easily serve as a vocabulary of motive in which individuals become convinced that they have no capacity for self-control and no responsibility for the consequences of their actions. How on earth can people control something that they truly believe is relentless and uncontrollable? Why would they even try? Regardless of how it is defined, addiction is not necessarily the same thing as danger: A substance that is nonaddictive can still be used destructively and harmfully, and an addictive substance can still be used safely and responsibly (Franklin, 1996).

The Unique and Variable Nature
of Human Drug-Taking Experiences

The addiction model fails to account for the diversity, the uniqueness, and the ineradicable social nature of human drug-taking experiences. Humans may use drugs because they expect them to bring them pleasure, to eliminate psychic discomfort, or to interrupt the onset of physical pain and discomfort (Grilly, 1994, pp. 118-121). Humans may even have a *psychogenic* addiction such that they believe that they are addicted and must continue drug use, but no physical basis for their addiction exists; it is only in their minds (Inciardi, 1992, p. 92).

Only humans use drugs because of peer pressure or because they want to be part of the drug-taking scene. In a drug subculture, users have the opportunity to develop new identities, relationships, lifestyles, and ideologies, all centered on drug use (Anderson, 1995, p. 366). Users will probably find others who are like them, and they may find this connectedness the most "intoxicating" part of their drug experience. Only humans can self-consciously and willfully discontinue their drug use because drug involvement becomes unattractive to them or inconsistent with other self-images or values that they have developed (Peele, 1985, p. 95).

In an informative book called *Heavy Drinking,* Herbert Fingarette (1988) explored the human dimension to drug use as he critiqued the claim that alcoholism is a disease. Though he concentrated exclusively on the drug of alcohol, his conceptualization of the drug-taking experience has great value for us in understanding *any* human drug experiences. Heavy drinking and alcoholism are not really physical disorders, he claimed. They are labels that are attached to cover a wide array of social and individual problems that are the result of a complex and poorly understood interrelationship between biological, psychological, social, and cultural factors (p. 27). The cardinal feature of alcoholism is its seeming uncontrollability and dysfunctionality. Reaching for that next drink without considering the complications seems to make no sense. Why would anyone sacrifice health, wealth, and happiness for what seems to be a transitory pleasure? A common explanation is that drug use can become a disease. Just as an individual with a cold cannot stop sneezing, an individual with an addiction supposedly cannot stop drinking, smoking, snorting, or injecting (p. 32).

The core of the "addiction model" of human drug use is the concept of craving, but craving is a poorly understood process. We have all experienced cravings of one sort or another (or believe that we have), but this hardly means that craving is the best or only way to conceptualize the dynamics of drug use (p. 41). Doubtless, some people, perhaps all people, can embark on some human activity such as drinking that will reach a point where they will throw caution to the winds and do something self-defeating. In these cases, it seems that a powerful momentum is at work and that no amount of persuasion, coercion, or self-control will be sufficient to create the level of self-restraint necessary to interrupt the bout of drug taking.

Why would someone drink to excess again and again? Fingarette (1988) insisted that heavy drinkers are not sick or diseased (or are no more so than anyone else). They are simply those people for whom alcohol drinking has become a central activity, not all that different from other central activities around which people organize their lives, such as jobs, hobbies, families, or community services (p. 100). When drinking alcohol evolves into a central activity, an individual's life is defined by drinking, and his or her major preoccupation becomes the pursuit of drink, drinking situations, and drinking companions. Central activities exert a great deal of power over us because they demand new relationships, new activities, new values, and new views of self and of others. They have a momentum of their own, and when one central activity becomes the hub of one's existence, it makes certain auxiliary activities far more likely and a host of other activities far less likely (p. 101). Just as any central activity reflects the characteristics of individuals involved in the activity—their physical abilities and traits, their personalities and temperaments, their social experiences—so does drinking. The defining elements of a heavy drinker's life are indicators of the character and quality of that life, not symptoms of some ill-defined disease (p. 107).

The social nature of human nature is no less apparent in persistent drug use than it is in other human experiences. People cannot change what they are and what they do entirely by themselves. A great deal more than good intentions and personal resolve is required to reconstruct one's life and to change central activities. However, it would be a mistake to discount the power and importance of will power in transforming one's central activities (pp. 109-111). Determination and dedication are important parts of any change—personal or social—and they are important in moving from one central activity to another or in shifting the priorities of one's central activities. We must not allow the tenets of the addiction model to blind us to the important role played by will, volition, and grit in human behavior.

It is not likely that the future holds some remarkable scientific breakthrough that will prove indisputably to all interested parties that drug use either is or is not an addiction. People who believe human addictions exist have no trouble finding the elements of an addiction in drug-taking behavior, just as they do in sexual behavior, violence, gambling, and overeating. People who believe that the term "addiction" fails to fit human drug-taking experiences have no trouble in finding inconsistencies and contradictions in the arguments of those people who do call human drug taking an addiction. One thing is certainly true: The word *addiction*—if it is reserved to describe some uniform, universal, biologically determined experience—is inadequate to account for the relativity and complexity of human drug taking.

The best thing might be simply to admit that terms such as "addiction," "addict," "dependency," "tolerance," "withdrawal symptoms," and "craving" are so mired in controversy and misunderstanding that they have no value and no purpose other than to defame and discredit; certainly, they contribute little to a better understanding of human drug use. Words such as "addiction" or "addict" do make it more likely that one will miss the huge impact of social, cultural, and personality factors on drug taking; they also make it more likely that one will see too much uniformity and consistency in drug-taking experiences. It may be best to say "no" to terms such as "addiction," "craving," and "addict" and to avoid them whenever possible (Alexander & van de Wijingaart, 1992).

Moral Entrepreneurs and Drug Crusades

Drugs are nothing more than inert substances until they are ingested by certain people who experience certain effects from them. What determines their moral status—good or bad—is how they are viewed and described. A drug of abuse at one time and place is not the same as it is at another, and groups that are despised and harassed for their drug use in one situation may be ignored or even praised in some other situation. Because images of drugs and drug users do change over time, and because so many groups have vested interests in what the images are, we must understand something about the nature of the crusades against drugs and drug users.

A drug crusade will not be successful without broad public support (Ryan, 1994, p. 218). To achieve this, drug crusaders must have sufficient time, influence, and resources to define a chemical substance and its use as problematic; convince others that what the crusaders say is true; devise a cure for the problem that is affordable, effective, and acceptable to others; and overcome any resistance to the crusade offered by other crusaders (Tuggle & Holmes, 1997, p. 80). Moral crusaders may subtly—and not so subtly—create new kinds of deviancy and new profiles of deviants by hunting, catching, and sanctioning people who had never been punished before (Victor, 1994, pp. 308-309).

The support of a budding moral crusade by scientific, medical, religious, or legal authorities is invaluable because these people are in the best possible position to explain, to support, and to legitimate the claims of the moral crusaders. Politicians, state representatives, and media spokespersons play a central role in creating and legitimating wars on drugs and the corresponding policies, messages, and images (Wysong, Aniskiewicz, & Wright, 1994, p. 461). In fact, without the involvement of respected authorities, most moral crusades will sputter and die (Victor, 1994, p. 310). These respected authorities may be responsible for initiating the moral crusades themselves.

Wars on drugs are usually legitimated on the grounds that they, like their military counterparts, can be won only if the cause is just and the soldiers are strong and determined. The major outcome of drug wars, however, is little more than increased social control over subordinate groups and greater attention to their patterns of drug use (Fagan, 1994). Drug crusades are almost always launched against groups of people who are already defined as a problem and a threat, and their patterns of drug use are used to justify an attack on them (Reinarman, 1996). Simply, they are being persecuted for being members of a despised or threatening group (Szasz, 1985, p. 65). The strangest irony of all is that the war against drugs may do greater damage to the social fabric and ruin more lives than the drugs themselves ever could (Trebach & Inciardi, 1993, p. 42).

U.S. history is filled with examples of drug crusades that boiled over into public hysteria (Musto, 1987). Alcohol, cocaine, cigarettes, opium, morphine, and heroin have all been the targets of moral reformers in the past, and these chemical substances and those who use them continue to be the object of periodic concern and attack (Brandt, 1991). Any usable chemical substance can be viewed as abusable by some group somewhere, and its members then set about to make things right. People with long enough memories can usually recall when today's "most dangerous substance on earth" was at one time proclaimed to be a safe and effective corrective for some human affliction (Inciardi, 1992). Before their criminalization, the major psychoactive drugs—opium, cocaine, marijuana—were administered freely to get people to work harder and longer at dangerous and exhausting jobs. These drugs increased productivity because they were a reward for hard work, and they took some of the edge off the pain of back-breaking labor (Szasz, 1985, p. 75).

An excellent example of how a chemical substance can be demonized and its negative effects exaggerated, thereby producing a drug scare, is the portrayal of phencyclidine (PCP) in the U.S. media. Phencyclidine was originally used as an anesthetic-analgesic in surgery (it dulled pain while the patient remained conscious), but it was eventually abandoned for use on human subjects because of its side effects. It can produce hallucinations, disorientation, and a range of unpleasant or upsetting physical symptoms (vomiting, rashes, numbness, lack of coordination, and excitement). When mixed with other drugs, usually LSD, phencyclidine is known as angel dust.

Phencyclidine reached the zenith of its use in the United States in the 1970s. It rapidly earned a reputation as a "devil drug" and came to be viewed as a substance capable of precipitating ghastly acts of interpersonal violence and self-inflicted injury and death. Descriptions of the drug in the print media and portrayals of it on television relied heavily on horror stories, especially ones that reported PCP users were unusually prone to gouge out their own eyes. If a stronger image than this one exists, it is hard to know what it could possibly be. It shocks, scares, repels, and creates a sense of urgency, while conveying an image of a drug user who is insane, antisocial, and totally out of control.

> Every new drug experience in America is handled in a stereotyped fashion by the media. Emphasis is placed on individual tales of dangerous, criminal, or self-destructive behavior by the drug-crazed. The myth is newly erected and slightly embellished with each new drug. (Morgan & Kagan, 1995, p. 210)

Were the negative images of PCP and its users totally false? Probably not—drug crusades may contain a kernel of truth—but facts about drugs are not usually the driving force behind drug scares. A gruesome tale such as the gouged-eye story is told and retold so often that it comes to offer instant proof of the correctness of the fears of drug crusaders. Eye gouging is so meaningful that facts cannot constrain it (Morgan & Kagan, 1995, p. 210).

A horror story can sensationalize certain events and make new chemical substances and new patterns of use the object of heightened public concern. What it cannot do, however, is to ensure that a demonized drug is actually the one thing responsible for all the bad effects blamed on it. Some of the problems associated with drugs exist because of how a drug is taken, how much is taken, who is taking it, or what impurities get into the body along with the psychoactive substance. Drugs purchased on the street, where a user can never be sure about how much or even what substance is being taken, can be dangerous to users, and some people can do dangerous things under their influence. People use these substances at their own risk, and some of these users will come to a tragic end. However, objective facts about the dangers of drugs are not really what account for the success of a moral crusade.

Helmer (1975) argued that myth and deception are the core of U.S. drug policy (p. 8). Past crusades against drugs such as opium, marijuana, and cocaine were not only about the dangers—real or imagined—of psychoactive drugs. Antidrug crusades were perfect covers for relentless attacks on minorities such as the Chinese, Mexicans, and African Americans, ostensibly because these people were addicted to drugs and dangerous to themselves and to others. The primary objective of these drug crusades was actually the exclusion of U.S. minorities from social life and the eradication of any threats that they might pose to moral crusaders' interests (Helmer, 1975; Latimer & Goldberg, 1981; Musto, 1987; Szasz, 1985). A drug crusade generates new images of drug use and drug users, and these images become weapons that some groups use to battle and control other groups.

The crusade against crack cocaine (and other forms of cocaine), reaching its zenith in the 1980s, is one of the more recent drug scares and one of the more enduring. Social constructions of drugs and drug users have powerful influences on how people act, and these social constructions are flexible and ever-changing. The transformation of the image of cocaine use from a harmless, recreational activity to a dangerous, if not deadly, addiction was accompanied by shifts in the symbolization of both cocaine and its users (Scheibe, 1994, p. 209). Some understanding of the social context of this moral crusade against cocaine will illustrate some of the elements found in *all* crusades against *all* drugs.

The 1970s and 1980s were a perplexing time in the United States. U.S. social diversity and cultural heterogeneity were frightening to those people who viewed them as disorganizing forces. A constellation of problems was particularly alarming: family breakdown, economic decline, disorder in schools, premarital sexuality, illegitimacy, pornography, adultery, community deterioration, crime, violence, and America's loss of influence in world events. If the problems were abundant, the causes of them were not. Too many people had lost their moral compasses, turning on with drugs and thereby neglecting their social responsibilities and failing to honor the social contract. The sensitization to the dangers of crack cocaine—farfetched or not—provided a scapegoat for social and personal problems that seemed to have increased dramatically as the years had passed (Reinarman & Levine, 1989, p. 127). Reducing the use and abuse of drugs seemed to be a particularly important part of the overall plan to rebuild the United States.

Anyone who was not enthusiastically and vocally against drugs was marked as being for them (as if only two options existed). The crusade against cocaine had great political and social utility for its promoters. In one fell swoop, it allowed them to show concern for the young, the old, and everybody in between. Being soft on drugs was tantamount to being soft on all that made America strong and healthy. Few politicians were in a strong enough position to be able to preach the virtue of tolerance in regard to people's drug-taking habits (Trebach, 1987). Fewer and fewer people from official society were willing to speak on behalf of cocaine (Erickson, Adlaf, Murray, & Smart, 1987).

One claim that received strong acclaim in scientific writings and the popular media is that cocaine poses exceptional threats to a user's health and well-being. Leading authorities asserted with confidence and usually with great fanfare that cocaine was too dangerous to be tried even once and that it posed extraordinary threats to any user. These claims, especially because they seemed to be beyond dispute, were enthusiastically embraced and used as one more bit of ammunition in the fight against drugs. The indisputable fact that certain drugs at high enough doses are deadly was transformed into a more general and bogus indictment of *all* illicit drugs as unusually dangerous.

Cocaine is a heart stimulant, so any imprudent or recreational use is cause for concern. However, *moderate* use of cocaine seems to be no more cardiotoxic than other stimulating activities, and the claim that no safe dose of cocaine exists is false (Alexander & Wong, 1990, p. 260). A link does seem to exist between *prolonged* and *heavy* use of cocaine and cardiotoxicity, but even here we must be cautious if truth is our objective. If a heavy user were to die from a heart attack while using cocaine—a rare event—it still would not prove that cocaine was the cause. The death might have been an atypical reaction to the drug, comparable to the deaths some people experience from using aspirin (Alexander & Wong, 1990, p. 260). Or it might have been due to an already existing health problem such as high blood pressure or arteriosclerosis. If this person had tried to jog or have sex or had gotten too excited over a sporting event, the outcome would have been the same (Alexander & Wong, 1990, p. 260). Another possibility is that it was not the cocaine itself but some adulterant in the street variety of the drug or some other drug used along with cocaine that caused the death (Alexander & Wong, 1990, p. 260).

A prominent scare story that has been associated with cocaine is the claim that large numbers of children are born addicted because of their mothers' use of the drug (Goode & Ben-Yehuda, 1994, pp. 216-219). The initial studies of babies who were born to mothers who used cocaine during pregnancy were not at all encouraging. They reported that these infants were more likely to be born prematurely; have a significantly lower birth weight; have smaller heads; suffer seizures; have genital and urinary tract abnormalities; suffer poor motor ability; have brain lesions; and exhibit behavioral aberrations such as impulsivity, moodiness, and lower sensitivity to environmental stimuli (Goode & Ben-Yehuda, 1994, p. 216). Findings such as these were picked up quickly by the media and just as quickly transmitted to the general public.

Enough evidence had accumulated by the 1990s to suggest strongly that the crack-baby syndrome is a medical fiction (Goode & Ben-Yehuda, 1994, p. 217). Mothers who smoke crack are more likely to engage in a variety of practices that strongly correlate with poor infant health, such as smoking tobacco, drinking alcohol, using other psychoactive drugs, refusing to exercise, following poor eating habits, failing to visit physicians regularly, and maintaining a generally reckless lifestyle.

> With crack babies, what we saw was [sic] pathological conditions *associated with* the use of cocaine that was [sic] automatically *assumed to have been caused* by the drug which later, careful research indicated, were in fact caused by very conventional conditions *about which there was very little subjective concern*. (Goode & Ben-Yehuda, 1994, p. 216)

Many things work together to determine the health of a newborn, and a mother's use of crack cocaine is just one of them.

Not only were helpless infants defined as an at-risk population, but so were the hale and the hardy. Growing attention was directed in the 1980s to the actual or imagined—it is impossible to know for sure which—use of illicit drugs by professional athletes (Leiber, Jamieson, & Krohn, 1993). Their drug use became the focus of attention precisely when difficult contract negotiations were under-way. Players were mainly concerned with issues of agency and pension plans, but management artfully focused most of the attention on illicit drug use among players. The players' wish to be free from the intrusion of drug testing was branded by the owners as an example of players' selfishness and irresponsibility. Why not go along with the plan unless something was being hidden? Media representatives tended to side with owners, and the players were portrayed as bad and generally ignorant of how damaging their drug abuse was to themselves and to the game. This realignment of interested parties increased the power of the owners and effectively undercut players' demands for higher salaries and more control over the game and even over their own personal careers.

Any evidence of an increase in drug use, no matter how slight, or any evidence of a shift in drug use patterns from one group to another is seized on and dramatized by the U.S. press (Lotz, 1991). The print media created a cocaine drug scare in the summer of 1986 by using "shocking numbers and graphic accounts" to make drug use in the United States appear ominous and widespread. These media representatives took raw data from surveys of drug use among high school seniors and young adults and fiddled with the numbers until they could be used to prove the existence of a drug epidemic.

> We found ample evidence of media workers snatching at shocking numbers from ISR [Institute for Social Research at the University of Michigan] press releases, smothering reports of stable or decreasing use under more ominous headlines, and distorting the cocaine problem to epidemic proportions as high as 40 percent of high school seniors. (Orcutt & Turner, 1993, p. 203)

When cocaine use could no longer be portrayed as an epidemic—drug stories had disappeared from all the major news magazines by the end of 1986—media magicians simply produced a new plague from the ashes of the old. *Newsweek* reported in the February 3, 1992, issue that LSD use was "rising alarmingly" among U.S. teens (quoted in Orcutt & Turner, 1993, p. 201). Like previous media claims, this one was mostly smoke and mirrors. The "alarming rise" in LSD use from 1989 to 1990 was countered by a much larger *drop* in cocaine use, but little was made of this decrease (Orcutt & Turner, 1993, p. 201). It seems that decreases in drug use are not nearly as newsworthy as increases.

The demonization of drugs and the persecution of drug users, curiously enough, are not intended to create a drug-free world. The war *against* certain drugs is almost always a war *for* other drugs (or at least for a particular social order where certain kinds of drug use are permitted or even encouraged).

> The abuse of tobacco, alcohol, and prescription drugs leads to far more serious consequences in terms of public health, violent behavior, spiritual deterioration, family disruption, death and disease—consequences that are all but ignored in the quest for a so-called drug-free culture. The drug-free culture being pursued is in fact a culture which allows citizens to consume culturally integrated drugs but excludes the drugs of other cultures. (Johns & Borrero, 1991, p. 79)

Representatives of the pharmaceutical industry do more than push drugs legally; they push a faith in the power of drugs to cure the afflictions of human existence by constantly reinforcing the message that the solution to personal problems will be found in a tablet, capsule, syrup, or injection. The drug pushing of the pharmaceuticals has serious consequences: death and injury to consumers from their use of inappropriate or dangerous drugs and a colossal waste of patients' money and national health resources (Chetley & Mintzes, 1992, p. 35). Drug companies do all they can to suppress reports of dangerous drugs or test results unfavorable to their economic interests. If this concealment fails, the results are whitewashed, and every effort is made to direct attention to the dangers of illegal drugs, rather than to the dangers of those drugs that are an important source of company profits (Clinard & Yeager, 1980).

Summary and Conclusions

Though animals regardless of species are sensitive to certain chemical sub-stances, it is the cognitive and symbolic abilities of humans and their complex and intricate social lives that are responsible for their drug-taking experiences. Ongoing disputes over fundamental questions are real and important: What is a drug? What is addiction? What is abuse? These questions are at the foundation of the interpretation of drug experiences. Some groups advocate using an addiction terminology to interpret the use of drugs; other groups advocate staying clear of words such as "addiction" and "addict" because they believe that these terms are used more to defame than to explain. Human drug experiences are uniquely human and show a great deal of relativity. Understandings about drugs and drug use are forged in the context of struggles between different groups, some of which take charge of interpreting social reality for everybody else.

References

Acker, C. (1991). Social meanings of disease: Changing concepts of addiction in the twentieth century. *Magazine of History, 6,* 28-29.

Alexander, B., & Wijingaart, G. F. van de. (1992). The disease of addiction: It's sick and tired. In A. Trebach & K. Zeese (Eds.), *Strategies for change* (pp. 275-279). Washington, DC: Drug Policy Foundation.

Alexander, B. K., & Wong, L. S. (1990). Adverse effects of cocaine on the heart: A critical review. In A. Trebach & K. Zeese (Eds.), *The great issues of drug policy* (pp. 257-267). Washington, DC: Drug Policy Foundation.

Anderson, T. (1995). Toward a preliminary macro theory of drug addiction. *Deviant Behavior, 16,* 353-372.

Becker, H. (1967). History, culture and subjective experience: An exploration of the social bases of drug-induced experiences. *Journal of Health and Social Behavior, 7,* 163-176.

Brandt, A. (1991). Up in smoke: How cigarettes came to be a controlled substance. *Magazine of History, 6,* 22-24.

Carnes, P. (1991). *Don't call it love: Recovery from sexual addiction.* New York: Bantam.

Chetley, A., & Mintzes, B. (Eds.). (1992). *Promoting health or pushing drugs? A critical examination of marketing of pharmaceuticals.* Amsterdam: Health Action International.

Clinard, M., & Yeager, P. (1980). *Corporate crime.* New York: Free Press.

Erickson, P., Adlaf, E., Murray, G., & Smart, R. (1987). *The steel drug.* Lexington, MA: Lexington.

Fagan, J. (1994). Do criminal sanctions deter drug crimes. In D. L. MacKenzie & C. Uchida (Eds.), *Drugs and crime: Evaluating public policy initiatives* (pp. 188-214). Thousand Oaks, CA: Sage.

Fingarette, H. (1988). *Heavy drinking: The myth of alcoholism as a disease.* Berkeley: University of California Press.

Franklin, D. (1996). Hooked/not hooked: Why isn't everyone an addict? In H. Wilson (Ed.), *Drugs, society, and behavior 96/97* (11th ed., pp. 54-65). Guilford, CT: Dushkin.

Goode, E., & Ben-Yehuda, N. (1994). *Moral panics: The social construction of deviance.* Cambridge, MA: Blackwell.

Grilly, D. (1994). *Drugs and human behavior* (2nd ed.). Boston: Allyn & Bacon.

Helmer, J. (1975). *Drugs and minority oppression.* New York: Seabury.

Henkin, W. (1996). The myth of sexual addiction. In R. Francoeur (Ed.), *Taking sides: Clashing views on controversial issues in human sexuality* (5th ed., pp. 58-65). Guilford, CT: Dushkin.

Inciardi, J. (1992). *The war on drugs II.* Mountain View, CA: Mayfield.

Johns, C. J., & Borrero J. M. (1991). The war on drugs: Nothing succeeds like failure. In G. Barak (Ed.), *Crimes by the capitalist state* (pp. 67-100). Albany: State University of New York Press.

Latimer, D., & Goldberg, J. (1981). *Flowers in the blood: The story of opium.* New York: Franklin Watts.

Leiber, M., Jamieson, K., & Krohn, M. (1993). Newspaper reporting and the production of deviance: Drug use among professional athletes. *Deviant Behavior, 14,* 317-339.

Levinthal, C. (1996). *Drugs, behavior, and modern society.* Boston: Allyn & Bacon.

Lotz, R. E. (1991). *Crime and the American press.* New York: Praeger.

Matveychuk, W. (1986). The social construction of drug definitions and drug experience. In P. Park & W. Matveychuk (Eds.), *Culture and politics of drugs* (pp. 7-12). Dubuque, IA: Kendall/Hunt.

McCaghy, C., & Capron, T. (1997). *Deviant behavior* (4th ed.). Boston: Allyn & Bacon.

Montagne, M. (1988). The metaphorical nature of drugs and drug taking. *Social Science and Medicine, 26,* 417-424.

Morgan, J., & Kagan, D. (1995). The dusting of America: The image of phencyclidine (PCP) in the popular media. In J. Inciardi & K. McElrath (Eds.), *The American drug scene: An anthology* (pp. 204-213). Los Angeles: Roxbury.

Musto, D. (1987). *The American disease: Origins of narcotic control* (expanded ed.). New York: Oxford University Press.

Orcutt, J. (1996). Deviance as situated phenomenon: Variations in the social interpretation of marijuana and alcohol use. In H. Pontell (Ed.), *Social deviance* (2nd ed., pp. 215-222). Upper Saddle River, NJ: Prentice Hall.

Orcutt, J., & Turner, J. B. (1993). Shocking numbers and graphic accounts: Quantified images of drug problems in the print media. *Social Problems, 40,* 190-206.

Peele, S. (1985). *The meaning of addiction: Compulsive experience and its interpretation.* Lexington, MA: D. C. Heath.

Ray, O., & Ksir, C. (1996). *Drugs, society, and human behavior* (7th ed.). St Louis: C. V. Mosby.

Reinarman, C. (1996). The social construction of drug scares. In E. Goode (Ed.), *Social deviance* (pp. 224-234). Boston: Allyn & Bacon.

Reinarman, C., & Levine, H. (1989). The crack attack: Politics and media in America's latest drug scare. In J. Best (Ed.), *Images of issues: Typifying contemporary social problems* (pp. 115-137). New York: Aldine.

Rice, R. (1998). The startling truth about sexual addiction. In L. Salinger (Ed.), *Deviant behavior 98/99* (3rd ed., pp. 217-219). Guilford, CT: Dushkin.

Ryan, K. (1994). Technicians and interpreters in moral crusades: The case of the drug courier profile. *Deviant Behavior, 15,* 217-240.

Scheibe, K. (1994). Cocaine careers: Historical and individual constructions. In T. Sarbin & J. Kitsuse (Eds.), *Constructing the social* (pp. 195-212). Thousand Oaks, CA: Sage.

Siegel, R. (1989). *Intoxication: Life in pursuit of artificial paradise.* New York: E. P. Dutton.

Staley, S. (1992). *Drug policy and the decline of American cities.* New Brunswick, NJ: Transaction.

Szasz, T. (1985). *Ceremonial chemistry: The ritual persecution of drugs, addicts, and pushers* (Rev. ed.). Holmes Beach, FL: Learning Publications.

Trebach, A. (1987). *The great drug war.* New York: Macmillan.

Trebach, A., & Inciardi, J. (1993). *Legalize it? Debating American drug policy.* Washington, DC: American University Press.

Tuggle, J., & Holmes, M. (1997). Blowing smoke: Status politics and the Shasta County smoking ban. *Deviant Behavior, 18,* 77-93.

Victor, J. (1994). Fundamentalist religion and the moral crusade against Satanism: The social construction of deviant behavior. *Deviant Behavior, 15,* 305-334.

Volkow, N. D., Wang, G.-J., Fischman, M. W., Foltin, R. W., Fowler, J. S., Abumrad, N. N., Vitkun, S., Logan, J., Gatley, S. J., Pappas, N., Hitzemann, R., & Shea, C. E. (1997). Relationship between subjective effects of cocaine and dopamine transporter occupancy. *Nature, 386,* 827-830.

Volkow, N. D., Wang, G.-J., Fowler, J. S., Logan, J., Gatley, S. J., Hitzemann, R., Chen, A. D., Dewey, S. L., & Pappas, N. (1997). Decreased striatal dopaminergic responsiveness in detoxified cocaine-dependent subjects. *Nature, 386,* 830-833.

Weil, A. (1972). *The natural mind.* Boston: Houghton Mifflin.

Wysong, E., Aniskiewicz, R., & Wright, D. (1994). Truth *and* Dare: Tracking drug education to graduation as a symbolic politics. *Social Problems, 41,* 448-472.

9

The Relativity of Mental Disorders

Introduction: Medicalization on 34th Street

A novel by Valentine Davies (1947) described what could have been the trial of the century: the time that Santa Claus was put on trial and his sanity questioned. The story really begins at the Maplewood Home for the Aged in Great Neck, Long Island. An elderly and kind gentleman, in the peak of health, whose name was Kris Kringle, spent his time making toys and smoking his pipe. One November, the physician at Maplewood (Dr. Pierce) gave Kris the bad news that he was being evicted. The reason was that state laws and Maplewood's charter allowed only people who were in good physical *and* mental health to live there. Kris's insistence that he was the one and only Santa Claus was a strong indication that he was mentally ill. Kris collected his belongings and left to make his way in the world, determined that he would not go to an asylum, no matter how bad things got. Luckily for him, he had a friend who lived and worked at the Central Park Zoo, and he made room for Kris at his place. It was a good arrangement because Kris had an uncanny ability with animals, especially with the reindeer.

Mr. Kringle wanted to see the Macy's Thanksgiving Day Parade. The big draw was that "Santa Claus" was escorted to his platform in the department store and started granting wishes to children of all ages. That year, the fake Santa that Macy's had hired was drunk, too drunk to carry out his duties. When Kris saw the man, he was furious. He complained directly to the personnel director and organizer of the parade, Doris Walker. She was visibly upset, dispatched the phony Santa, and asked Kris to please take his place. Kris consented only because he did not want to disappoint the children. He did so well that he was offered a job in Macy's as the permanent Santa, a position that he reluctantly accepted.

Kris ran afoul of the entrenched beliefs of other employees of Macy's by continuing to insist (at least when asked) that Santa Claus did exist and that he was the one and only. Questions arose about his sanity and his dangerousness. In fact, Ms. Walker wondered if Kris was as harmless as he appeared. She consulted Macy's only expert on psychology and vocational guidance, Mr. Albert Sawyer. He was an arrogant, dogmatic fellow who thought he knew practically everything about practically everything. He told Doris Walker that he would be pleased to interview Kris. Sawyer concluded without much to go on that Kris had a fixed delusion and could be dangerous, and he reported this to Ms. Walker.

That December, Sawyer was scheduled to give a lecture entitled "Exploding the Myth of Santa Claus." Kris attended because he believed that he had some valuable information to contribute. At the lecture, Sawyer vociferously condemned the myth of Santa Claus and insisted that any adult who believed in Santa Claus revealed himself or herself to have an incomplete and neurotic personality. Sawyer opined that the "vicious myth" of Santa Claus had done more harm in the world than had opium. Kris's presence unnerved Sawyer. Sawyer's speech became incoherent, and he garbled sentences. Finally, Sawyer demanded that the old "jackanapes" be removed from the room. Kris refused to move. Sawyer advanced on him, Kris raised his cane in protection, Sawyer grabbed it and tugged, Kris lost his grip, and Sawyer hit himself in the cheek. He screamed that he had been attacked by Kris. The altercation with Sawyer (along with Kris's claim that he was Santa Claus) worried others. Steps were taken to have a hearing to determine if Kris should be committed to an institution for his own good and for the safety of the people around him.

The commitment hearing was under the charge of Judge Henry X. Harper. A Mr. Mara represented the state of New York, and a Fred Gayley represented Kris Kringle. The first witness Mr. Mara called was the defendant, Mr. Kringle. Kris was a right sprightly old elf, and Judge Harper smiled when he saw him in spite of himself. Mara got right to the point and asked Kris if he believed that he was Santa Claus. "Of course!" was the reply. The courtroom was hushed. The old gentleman didn't even try to hide his craziness! Mara rested his case and sat down, confident that he had proved that the old man was insane.

Fred Gayley was desperate as he sat in court. He was trying to prove by a preponderance of the evidence that Kris Kringle was mentally healthy. Mara was reading reports into the record about inmates of state institutions who thought that they were Napoleon, Caruso, or Tarzan. Delusions like Mr. Kringle's were common enough. It looked as if Kris might be committed to Bellevue after all. However, a fortuitous turn of events kept this from happening. Gayley was summoned from the courtroom and was shown bundles and bundles of letters, each one addressed to Santa Claus. (Every letter addressed to Santa Claus that had been received at the local post office had been bundled up and delivered to Kris at the courthouse.) Struck by an idea, Gayley returned to the courtroom, and when his opportunity to speak came, he reminded the court that it was a serious offense to deliver mail to the wrong party. He then produced three letters from his jacket, each one addressed simply to "Santa Claus, U.S.A." Their delivery to Kris proved (or so Gayley claimed) that Kris was recognized by competent authority to be the one and only Santa Claus. The court was unimpressed: Three letters were hardly enough to make such a strong claim. But Mr. Gayley had further exhibits. Judge Harper insisted that they be put on his bench, Gayley complied, and the judge was eventually covered by a mountain of mail. Judge Harper ruled that if the United States of America believed Kris Kringle was Santa Claus, his court would not disagree. The case was dismissed. Kris was deeply moved. Smiling happily, he rushed to the bench and wished the judge a merry Christmas. Then Kris disappeared into the night. After all, it was Christmas Eve.

Davies' book about the fictional trial of Santa shows something factual: Mental illness is a social construction. Claiming to be Santa is no different in certain respects from claiming to be Judge Harper, Fred Gayley, or Mr. Mara. The rendering of a judgment that a mental disorder exists comes from the determination that an individual is not what he or she claims to be and that he or she should *not* be feeling or doing what he or she is feeling or doing. The dispute over the goodness or badness of Kris's so-called delusion, or even the determination that he had one, cannot be separated from the social context in which that decision is made. A pivotal issue seems to be whether "competent authority" will state for the record that an individual is sick, insane, or abnormal. Almost any one of us could find that his or her thoughts, actions, or feelings were being construed as indicators of a mental disorder by somebody.

The Social Nature of Mind

One of the best and most lucid accounts of "mind" was written by a social behaviorist named George Herbert Mead (who is now usually referred to as the father of symbolic interaction theory). Mead (1934) insisted that mind and brain are not the same *at all*. Brain is a biological organ, and without it, mind is impossible. However, the existence of a brain does not necessarily mean that an organism has a mind, and it is possible to have mental processes that may appear odd to others that are not caused by any brain *dysfunction* or abnormality. Thinking, mind, or reflexive intelligence—Mead used these terms interchangeably—is an inner conversation through the use of symbols or significant gestures (p. 47). The organization of brain processes through language is what mind is all about.

What does it mean to lose a mind? You can lose weight; you can lose your hair; you can lose an earring or a wallet; you can lose all your money; you can even lose your life. But how can you lose a mind? Psychiatrists have created a definition that makes "losing a mind" look objective, uniformly debilitating, and idiosyncratic. In the fourth edition of the *Diagnostic and Statistical Manual of Mental Disorders* (*DSM-IV*),

> [Each mental disorder is conceptualized as] a clinically significant behavioral or psychological syndrome or pattern that occurs in an individual and that is associated with present distress (e.g., a painful symptom) or disability (i.e., impairment in one or more important areas of functioning) or with a significantly increased risk of suffering death, pain, disability, or an important loss of freedom. (American Psychiatric Association [APA], 1994, p. xxi).

The authors of this definition stated that these syndromes or patterns absolutely *cannot* be an "expectable" and "culturally sanctioned" response to a particular event (p. xxi). They insisted further that the disorder must be a manifestation of a *dysfunction* in an individual (behavioral, psychological, or biological). They declared that neither deviant behavior (political, religious, or sexual) nor conflicts between an individual and society are mental disorders *unless* they are a symptom of some individual dysfunction (p. xxii). This conceptualization of mental illness creates a false separation between those people whose reactions make no sense and everybody else.

Psychiatrists' desire to separate deviance, interpersonal conflicts, and all the rest from individual dysfunctions is based on little more than wishful thinking. They really have no way to know an individual dysfunction when they spot one or to separate it from other "expectable" and "culturally sanctioned" responses (Fernando, 1991). *Mental disorder* is a vague, ambiguous, and confusing term. Human behaviors, thoughts, and feelings are just too diverse to be neatly separated into two categories of "normal" and "pathological."

Many people—both inside and outside of the mental health field—are skeptical that disorders can be differentiated from nondisorders and that mental problems can be neatly separated from problems that are physical, behavioral, social, or moral (Kirk & Kutchins, 1992, p. 226). In fact, the architects of *DSM-IV,* in a moment of unusual candor, confess as much:

> In DSM-IV, there is no assumption that each category of mental disorder is a completely discrete entity with absolute boundaries dividing it from other mental disorders or from no mental disorder. There is also no assumption that all individuals described as having the same mental disorder are alike in all important ways. (APA, 1994, p. xxii)

This is quite a remarkable statement! It acknowledges that mental health professionals really cannot separate mental disorders from each other or even from a condition called "no mental disorder." Individuals who have been given the same diagnostic label are not even necessarily alike in terms of the defining features of the diagnosis, let alone in terms of other biological, psychological, or social factors. If psychiatrists, who have the strongest incentive to formulate a definition of mental disorder that is beyond reproach, can do no better than they have done, then it probably cannot be done. Any effort to define *mental disorder* in some objective and uniform way is doomed to failure.

The Myth of Mental Illness

It seems to defy common sense, logic, and all that we know about social deviance to believe seriously that it is possible to distinguish the socially acceptable from the socially unacceptable in terms of some individual dysfunction. Finding mental illness is an ineradicable social process of deciding who is normal and who is not that has little to do with a diagnosed individual's physical characteristics (Turner & Edgley, 1996, p. 434). What wayward chemical or brain pathology could possibly exist to differentiate those individuals who hear socially *acceptable* voices such as God's or Allah's from those individuals who hear the socially *un*acceptable voices of space invaders or some dead relative? It is not really hearing voices that is the problem but whose voice is heard (or where, or when, or why) (Turner & Edgley, 1996, p. 437). What wayward chemical or brain pathology could possibly distinguish the socially *acceptable* pretending of children, adolescents, actors, magicians, defense attorneys, lovers, parents, flatterers, and inveiglers from the socially *unacceptable* pretending of people who claim to be (and may believe that they are) Napoleon, Jesus Christ, or Santa Claus? It is easy to see how a society could make clear distinctions between the sick and the well or between the acceptable and the unacceptable, but it is much more difficult to see how neurochemistry could (Turner & Edgley, 1996).

Szasz (1973) nicely illustrated the variable and relative nature of mental illness while showing the folly of searching for invariant biological causes of it:

> If you believe that you are Jesus, or have discovered a cure for cancer (and have not), or the Communists are after you (and they are not)—then your beliefs are likely to be regarded as symptoms of schizophrenia. But if you believe that the Jews are the Chosen People, or that Jesus was the Son of God, or that Communism is the only scientifically and morally correct form of government—then your beliefs are likely to be regarded as reflections of who you are: Jew, Christian, Communist. This is why I think that we will discover the chemical cause of schizophrenia when we will discover the chemical cause of Judaism, Christianity, and Communism. No sooner and no later. (pp. 101-102)

Sharp lines between the normal and the abnormal cannot be drawn in the murky world of right/wrong, legal/illegal, proper/improper, healthy/sick, moral/immoral—that is in the complex and ever-changing system of rules and rule breaking—and when lines are drawn, they have more to do with sociocultural settings, languages, meanings, and interpersonal conflicts than with an individual's body chemistry or brain structure. Any psychological disorder tells us not only about the anguish of an individual but also about patterns of interaction, what they are, and how they have unfolded over time (Banton, Clifford, Frosh, Lousada, & Rosenthall, 1985, p. 74).

When we speak of a sick mind, Szasz (1973) insisted, we can only be speaking metaphorically. To say that a person's mind is sick is like saying that the economy is sick or that a joke is sick (p. 97). A metaphor is a figure of speech wherein two things are contrasted to present a new way of thinking about an old concept. "The moon is a ghostly galleon" and "she has abdominal muscles of steel" are both metaphors. These words paint pictures that would be impossible without them, but we must not confuse a metaphor with a scientific description or a truth statement. The moon is not a boat, and no human can have abdominal muscles made of metal. When a metaphor is mistaken for reality and is used to advance some social interests and impede others, then we have the makings of myth (Szasz, 1973, p. 97).

Declarations that an individual has a mental illness, rather than being based on sophisticated knowledge of disease, are really judgments about one individual's personal, social, and ethical problems in living, offered by others who do not share or subscribe to the same understandings (Szasz, 1974, p. 262). The basic struggle for survival in the human kingdom is a struggle over *words*: Define or be defined (Szasz, 1973, p. 20). The only way that a judgment could be made that an individual who claims to be Santa Claus (or to be Jesus Christ or to be hounded by the communists) is mentally sick is for a definer to conclude that this claim is false by comparing it to other claims that he or she believes are healthier and happier. Feeble minds can be understood only in the context of a matrix of relationships in which certain pursuits become so central—material success, upward mobility, or intellectual superiority—that anyone who does not display the appropriate levels of self-interest, emotional maturity, and cognitive ability will be branded as mentally defective (Trent, 1994).

Troubles and tensions are regular features of human relationships, and life is an ongoing struggle to maintain self-respect and personal worth in the face of continual assaults on them. The label of mental illness is used to make inter-personal conflicts and problems in living easier to manage by diagnosing them as individual dysfunctions that can be cured through the marvels of medical science. What adult would not prefer to believe that a child's misbehavior sprang from some defect in the child rather than from some defect in how the child was socialized? Biochemical explanations of mental illness may seem attractive because they release individuals from any personal responsibility for their own problems in living (or for the suffering of others) and from any obligation either to repair or to leave their strained relationships (Breggin, 1991, p. 94). Our adversary, if one exists, is not some ill-defined disease of the mind but a general malaise that interferes with the development of rewarding and harmonious social relationships (Szasz, 1960). So-called mental disorders may be very reasonable responses to very unreasonable or complicated social situations (Umberson, Chen, House, Hopkins, & Slaten, 1996, pp. 852-854).

The Dominion of Psychiatry:
Mental Health Buccaneers and Social Control

Psychiatric Tyranny and the Manufacture of Madness

Deviance does not float down from the skies, automatically attaching to deviants and automatically missing nondeviants. For deviance and deviants to exist as meaningful categories, somebody has to judge, to portray, to stigmatize, to insult, to abuse, to exclude, or to reject (Sumner, 1994, p. 223). Though many groups and professions benefit from a societywide interest in being mentally healthy (Herman, 1995), psychiatry sits at the top of the heap (Dowbiggin, 1991). This medical specialty has a near-monopoly over definitions of mental health and mental illness, and its practitioners have a disproportionate influence over the treatment and cure of mental disorders. It is impossible to talk about mental illness without discussing the structure and function of psychiatry and the central role that it plays in the manufacture of madness.

It is an article of faith for most psychiatrists that a brain and/or chemical dysfunction will be uncovered for all mental illnesses. The search for biochemical causes of mental disorders is what allows psychiatrists to stay at the top of the mental health hierarchy, and it legitimates psychiatry as a medical specialty (Breggin, 1991, p. 23). It allows psychiatrists to claim that they, just like other physicians, specialize in the discovery and treatment of pathological body structures that just happen to cause diseased minds. When psychiatrists were unable to find structural abnormalities in the body and trace them to particular mental diseases—the usual situation—they were not particularly disheartened. They simply *invented* mental diseases through the simple but effective strategy of proclaiming that some things were sick and in need of treatment, hoping that they would eventually find some biochemical abnormality to go along with each newly named disease (Szasz, 1974, p. 12).

Critics of psychiatry have deplored the arbitrary and pseudoscientific way that psychiatrists determine what is sick and what is not. Armstrong (1993) took psychiatrists to task for using what she called the Peter Pan method to create and validate diagnostic categories: Clap if you believe in fairies (p. 152). In this case, the "fairies" are things such as schizophrenia, premenstrual dysphoric disorder, simple phobia, and post-traumatic stress disorder. Only the hopelessly naive or extremely gullible could believe that high consensus among psychiatrists is the same thing as the scientific discovery of mental illnesses. Breggin (1991), himself a psychiatrist, also took psychiatrists to task: "Only in psychiatry is the existence of physical disease determined by APA presidential proclamations, by committee decisions, and even, at times, by a vote of the members of APA, not to mention the courts" (p. 141). The judgments of psychiatrists are part of what they judge, and mental disorders are not separable from the language that names them (Kovel, 1988, p. 134). Modern psychiatry has achieved the enviable position such that proclamations are taken as truth, wishes can become achievements, and propaganda is taken as science (Breggin, 1991, p. 182).

Psychiatry is at a turning point. Its practitioners can abandon their traditional concern with the mind and specialize in the treatment of the brain. If they do this, they will be indistinguishable from neurologists. Or they can start worrying more about mind and leave the brain alone. If they choose that path, they will be little different from nonmedical therapists and counselors. In recent years, psychiatrists have tried to align themselves more closely with specialists in brain abnormalities (Szasz, 1985, p. 712). They have labored to convince the public and the federal government that a shortage of psychiatrists exists and to convince health insurers that psychiatry is a bona fide medical science.

> [Organized psychiatry] develops media relationships, hires PR firms, develops its medical image, holds press conferences to publicize its products, lobbies on behalf of its interests, and issues "scientific" reports that protect its members from malpractice suits by lending legitimacy to brain-damaging technologies. (Breggin, 1991, pp. 366-367)

Some therapists will admit that good psychiatry is an art more than a science, requiring skills not taught in medical school—interpersonal skills that have little to do with drugs, science, or medicine—but this admission, though true, does blur the line between science and art, between healer and healed, and between medicine and religion/ethics (Neill, 1995, p. 219).

What are psychiatrists? Are they brain doctors, mind doctors, body doctors, friends and confidants, sages, healers, spiritual advisers, philosophers, entrepreneurs, cops in lab coats, or what? The identity crisis in psychiatry has far-reaching and often harmful effects. Former patients—they call themselves "psychiatric survivors"—regularly protest at APA meetings to publicize what they believe are injuries inflicted by their treatments.

Psychiatry and Social Control

Psychiatric intervention is used to handle potentially unruly or errant populations such as the insane, the deranged, the idle, the ignorant, or even the deprived. The psychiatrization of more and more problems in living makes it quite likely that manifestations of social problems are being transformed into individual problems of adjustment (Szasz, 1997, p. 215). Caplan (1995) provided an excellent example of how this can work. At a clinic where she was employed early in her career, she was told by her supervisor that her job was *not* to attempt to find better jobs or additional welfare monies for the poor families who came to the clinic for help (even though that is what Caplan believed they really needed). Her charge was to find some appropriate psychiatric label that would fit her clients' complaints and then to suggest some appropriate medical treatment—such as psychotherapy—that was an established part of traditional mental health care (p. 280). When medical treatments are adopted as the principal way to soothe all the people who are hurt by inequality, unemployment, poverty, divorce, racism, sexism, or other social conditions, it becomes less likely that people will work together to change those things that are really responsible for much of their unhappiness.

One of the populations that psychiatry has traditionally controlled is children, almost always under the guise of helping. The psych deck, however, is already stacked against them. Not only are they young and powerless, but they are defined as having individual dysfunctions, and they have probably alienated central figures in their families, their schools, and their communities. The claim that some children have serious troubles, though certainly true, can serve as a convenient smokescreen.

> Poverty was let off the hook. Social injustices were let off the hook. Parents were let off the hook. Lousy schools and dysfunctional teachers were let off the hook. There was simply something biologically wrong with these children that accounted for all the things that teachers, parents, and Boy Scout leaders did not like about a whole panoply of childhood behaviors: not sitting still, not paying attention, not learning to read correctly (on time), butting in. (Armstrong, 1993, p. 196)

It is impossible for psychiatry to determine how many children's problems in living are individual dysfunctions and how many are actually normal responses to abnormal situations. Psychiatrists cannot even agree on exactly *who* would benefit from psychiatric treatment. Are we talking about behavioral misfits, the emotionally ruined, the angry, the unmotivated, the irresponsible, the intractable, the socially deprived, the abused, the maladjusted, the promiscuous, the frigid, the rigid, the amoral, the immoral, the insane, the depressed, the hyperactive, the diseased, all of the above, none of the above, or what?

What gets a child defined as mentally ill is almost always some persistent conflict or disagreement with some authority figure. Most of these children usually have little difficulty getting along with their peers in situations of their own making. To call a lack of fit between children and adults a mental illness of the child is at best a distortion and at worst an outright lie. Too much is blamed on the child and not enough on the people and situations that he or she is supposed to manage successfully. Practically any characteristic of children that irritates, interferes with, or frightens authority figures can be medicalized and used to justify psychiatric intervention into their lives. A child's inappropriateness may be little more than a method of adapting to an intolerable or undecipherable situation (Breggin & Breggin, 1994, p. 114).

Sense and Nonsense in the Psychiatric Enterprise

The Crafting of Psychiatric Nosology

A script for human experience, especially if it is elaborate, agreed on, widely shared, formalized, and viewed as divine, sacred, or inviolable, gives the impression that certain experiences are far more reasonable, healthy, necessary, uniform, and universal than they really are. Unsurprisingly, people who have taken a great deal of time and trouble to master a particular script—learning its language, coming to believe in its importance and value, appreciating its logic—have trouble accepting that it is just one script among many. People walk to a different beat because they hear a different drummer, but ideal scripts tend to make it appear that only one march exists. Psychiatric nomenclature, along with the medicalization of problems in living that it justifies, is characterized by a great deal of ethnocentrism and the imposition of standards of health and normality on people for whom they are woefully inadequate (Bartholomew, 1997).

One of the more monumental ideal scripts for classifying human acts, thoughts, and feelings is the DSM, currently the *DSM-IV* (APA, 1994). It is the largest and most inclusive of all the *DSM*s (there have been three previous editions plus a revision of the third edition). It is 886 pages long, has 16 major diagnostic categories, and describes almost 400 separate mental disorders. The only thing that these approximately 400 different disorders have in common is that they all appear in *DSM-IV*; any other similarity is purely coincidental.

The architects of the *DSM* claim that logic, science, and well-conducted research have all worked together to tell them what is normal and what is abnormal. However, the *DSM* has little to do with science; in fact, according to Caplan (1995), it is "shockingly unscientific" (p. 31). The contents of the manual are determined primarily by the gatekeeping efforts of the small number of influential psychiatrists who have the directive to decide which disorders will be allowed to appear and which will not (p. 185). This document does not solve the mystery of why people are mentally ill because it was constructed primarily to benefit mental health professionals, not their patients (Kovel, 1988). A handbook of mental disorders has many benefits for psychiatrists (Kirk & Kutchins, 1992, p. 219). It gives them a sense of competency and control; it gives them some professional respectability; and it gives them some legitimacy because if psychiatrists cannot reliably and accurately diagnose a patient's mental illness, how can they treat it? Of course, it also allows them to receive compensation for their services from medical insurers.

The claim that the *DSM* provides a uniform and objective scheme for classifying human mental disorders is unfounded. As the *DSM* has gone through its many revisions, it has expanded its list of mental disorders and lost both its sense of proportion and its utility as a guidebook that could reasonably help people to overcome some debilitating human afflictions.

> A comprehensive medically oriented diagnostic manual could be helpful for some purposes, but only if it were much narrower in scope, included only clearly distinguishable mental disorders that entail severe consequences and didn't pander to insurers, drug companies and therapists by medicalizing so many social problems. (Kirk & Kutchins, 1994, p. A17)

If architects of the *DSM* had packaged their document honestly—as a sourcebook of diagnoses based on a sampling of studies of unknown validity and reliability, as well as on ethics, social values, hunches of psychiatrists, psychologists, and others in the mental health fields, and reflecting political pressures, tradition, and economic/business decisions—it could be honestly accepted or rejected for what it is. As it stands now, it is an ailing document. Though reasonable people can disagree over what should or should not be included in a manual of mental disorders, they cannot reasonably disagree that the definitions of terms in *DSM* are fuzzy and that people—therapists and nontherapists—use them without much precision. The only restrictions on what appears in the manual are internal negotiations between members of the APA and the desire of psychiatrists to avoid outside ridicule and interference (Kirk & Kutchins, 1992, pp. 185-186).

Some of the general silliness of *DSM-IV,* and a primary reason to be suspicious of its implied claim of uniformity and universality, can be appreciated by examining a few of the outlandish disorders that appear on its pages. One of these is the "disorder of written expression" (APA, 1994, pp. 51-53). The essential feature of this disorder is that a person's writing skills (as measured by standardized tests) are "substantially below" those expected given the individual's chronological age, measured intelligence, and age-appropriate education. The presence of this disorder is indicated by things such as grammatical or punctuation errors within sentences, poor paragraph organization, multiple spelling errors, and excessively poor handwriting. What is wrong with this picture? Though a few students may be bad writers because they have sick minds, the manual provides no clue about how to separate them from those who are poor writers because they were poorly trained or were unresponsive to writing instruction. "Mathematics disorder" (APA, 1994, pp. 50-51) covers individuals who are unable to learn multiplication tables, to carry numbers correctly, to decode word problems, to obey mathematical signs, or to copy numbers or figures correctly. Once again, something that can have a multitude of causes (and is really only a problem in places where math skills are necessary for daily living or to succeed in school) is transformed into an individual pathology through the process of naming and proclaiming. Perhaps the silliest of the silly is "learning disorder not otherwise specified." This diagnosis covers learning disorders that do not meet the criteria for learning disorders:

> This category might include problems in all three areas (reading, mathematics, written expression) that together significantly interfere with academic achievement even though performance on tests measuring each individual skill is not substantially below that expected given the person's chronological age, measured intelligence, and age-appropriate education. (APA, 1994, p. 53)

Practically any problems relating to learning could be pigeonholed here, regardless of test scores, any of which could be average, above average, or even far above average. The APA seems to have covered all its bases.

The *DSM-IV* includes Appendix B, a list of proposed disorders that require further study (APA, 1994, pp. 703-761). Each of the proposed disorders, the authors of *DSM-IV* claimed, was subjected to a careful empirical review, and wide commentary was solicited from the field (1994, p. 703). Each one apparently failed to receive the kind of support necessary to warrant making it a bona fide disorder in the body of the text. One wonders how one of these—caffeine withdrawal—was subjected to this "careful empirical review and wide commentary from the field." One also wonders why it is not a bona fide disorder in the body of the text. The diagnostic criteria of caffeine withdrawal are heavy caffeine use, some physical withdrawal symptoms from abrupt cessation of caffeine (headaches, marked fatigue or drowsiness, marked anxiety or depression, or nausea or vomiting), and clinically significant distress or impairment in social, occupational, or other important areas of functioning. How hard can it be to find someone who fits this profile? The most baffling thing, however, is that caffeine withdrawal *does* get included in the manual, but in a sneaky way. On page 215 is "caffeine-related disorder not otherwise specified," one of the substance-related disorders. This diagnosis is reserved for "disorders associated with the use of caffeine that are not classifiable as Caffeine Intoxication, Caffeine-Induced Anxiety Disorder, or Caffeine-Induced Sleep Disorder" (APA, 1994, p. 215). To illustrate this "not otherwise specified" disorder, the APA directs readers to caffeine withdrawal and its research criteria in Appendix B on page 708! As Alice said on her journey through the looking glass, "curiouser and curiouser."

The kind of disorders we have been examining—learning disorders or caffeine withdrawal—may be absurd, but we might expect some absurdities when new mental disorders are introduced with each revision of the manual. Still, we would expect some of the classic disorders such as schizophrenia to be the model of objectivity and scientific integrity. If not these, then what? However, we would be disappointed. Research into the causes and nature of schizophrenia is extensive but in many ways contradictory. A dazzling array of symptoms is exhibited by schizophrenics, and about 20 distinct explanations have been proposed to make sense of schizophrenia (Gallagher, 1995, p. 86). Heinrichs (1993) reviewed 70 years of research on schizophrenia. His principal conclusion is that very little is known or understood about it and that it suffers from a "heterogeneity problem"—a polite way of saying that the term is used in so many different ways to cover so many different things that it is losing much of its meaning. Whatever schizophrenia is, it is not pure abnormality, and, if it is a disease, it is one of the most peculiar around. As Straus (1991) commented, "Schizophrenia is very different from an illness like a broken leg. You cannot break a leg, take off the cast in order to play football, and then, after the game, put on the cast again and be an invalid" (p. 82). A person labeled schizophrenic can alternate back and forth between normality and abnormality on a moment-to-moment, hour-to-hour, or day-to-day basis (Strauss, 1991, p. 83). Schizophrenia has been branded as a defective, deficient, disrupted, and disorganized construct (Wiener, 1991, p. 205).

Meltdown in the *DSM*

The APA has occasionally messed with the wrong group and then found itself the target of unwanted attacks. Sometimes these conflicts were little more than skirmishes, but other times they were outright wars over the validity and reliability of some of the mental disorders that psychiatrists included in the manual. The gay community successfully pressed its claim that homosexuality is not an illness, and it was eventually omitted from psychiatric nomenclature (Bayer, 1981). No doubt, similar battles would have been fought over many more of the disorders in the manual if other offended groups had been able to muster the resources to challenge organized psychiatry's monopoly over determining what is normal and what is not. One group that the APA has consistently offended and enraged is women. Women's battles with psychiatry show that tension and discord are at the heart of any effort to pigeonhole behavior and people, especially if those people prefer to be left alone and have the resources and abilities to defend themselves. When psychiatrists took on women, they had a tigress by the tail.

In the *DSM*'s third, revised edition (*DSM-III-R*; APA, 1987), an appendix appeared that included three disorders that were classified as needing further study: sadistic personality disorder, self-defeating personality disorder, and late luteal phase dysphoric disorder (LLPDD). The first two were cast out of *DSM-IV*, but the last was renamed as premenstrual dysphoric disorder (PDD) and was included in Appendix B. Premenstrual dysphoric disorder essentially transforms women's more extreme premenstrual experiences—anxiety, depression, irritability, despondency, anger—into a mental disorder. The APA apparently does not really consider PDD a proposed diagnosis that requires further study (the logic of appearing in Appendix B) because it *does* appear in the body of the text as a "depressive disorder not otherwise specified" (APA, 1994, p. 350). A disorder that is relegated to a provisional appendix can apparently be called into service as a bona fide mental disorder when the need arises.

In the 1950s, Katherine Dalton, a British physician, concluded that women's raging hormones were responsible for their mood swings and antisocial behavior that became especially prevalent during their premenstrual periods. Dalton and an associate called this constellation of hormonally caused afflictions "premenstrual syndrome" or PMS (Francoeur, 1996, p. 44). PMS is an extension of a 19th-century belief that women were swayed by the sex hormones produced by their ovaries. These hormones were blamed for premenstrual tension, and premenstrual tension was blamed for women's apparent inability to work and to carry out other social responsibilities (Frank, 1931; Rittenhouse, 1991, p. 419).

Critics of the view that PMS is a clinical medical disorder believe that too much hoopla has been generated over far too little—a case of too much ado creating the very problems that the ado was supposed to correct in the first place. Even if some monthly changes can be found in premenstrual women, they may have more to do with social relationships, selective perception, and how women interpret their physical changes than they do with raging hormones and ovarian dysfunctions. Some women do experience a debilitating syndrome of maladies on a regular basis, but this can hardly be typical for most women or even for a substantial minority. No consistent biological markers have been found to separate PMS women from non-PMS women, and practically all women handle what they must handle—jobs, children, households, relationships—regardless of how they feel during their menstrual periods. A woman's mood swings seem to have more to do with the quality of her life than with the condition of her ovaries (Tavris, 1996, p. 53). Representatives of different social worlds or domains—

groups of people who share views, concerns, and commitments—continue to contest the status of PMS as a mental illness, and PMS continues to be a different construct to different audiences: women, health and mental health practitioners, and scientists (Figert, 1995, p. 58).

Premenstrual changes—physical, mental, emotional—may occur in women (to a greater or lesser degree), but these altered states are not necessarily mental disorders. Women who do report the most acute problems live in situations that are characterized by stress, poverty, insecurity, and abuse; the treatments that seem to help these women the most are participation in self-help groups, changes in diet, and increases in exercise (Caplan, 1995, p. 156). Women may feel depressed, angry, or irritable all the time but may notice it more during their periods or believe that it is more legitimate for them to give expression to these feelings when they are premenstrual (Caplan, 1995, p. 162). Women's growing independence and assertiveness, as well as their anger over persistent gender inequality, can be inauthenticated or delegitimized by being blamed on PMS. This social transformation of the political into the biological helps to "cool out" political tensions. Women may be less likely to vent their female fury—or to unite with others to identify joint problems and work to eradicate them—because they define the turbulence in their lives as physical, abnormal, and transitory (Laws, 1983).

Martin (1992) reported that PMS is branded as a serious physical problem that makes women unfit for most things at precisely those points in time when it is necessary to keep women out of the labor force and in the home. When women were needed to assist in the war effort, little attention was paid to PMS. However, when men returned from war and needed jobs, women were nudged back into more traditional roles by the discovery (or rediscovery) of the crippling impact that PMS had on them. When women's participation in the labor force is seen as a threat, menstruation is more likely to be constructed as a liability (pp. 120-121).

Summary and Conclusions

The efforts to define mental disorders in some uniform, systematic, and objective way—though understandable—are doomed to failure. Terms such as *mind* and *mental disorder* are just too fuzzy and unscientific to be used in objective ways to define and to differentiate among people for what they say, think, feel, and do. So-called mental disorders are constructed within the context of a society, and they tell all us more about the quality of life in that society than they do about individual dysfunctions. Sharp lines cannot be drawn between the sick and the well in the complex and ever-changing world in which we live.

Whatever good a manual such as the *DSM* does is hardly compensated for by the great harm that it does to our understanding and acceptance of human diversity. Mental illness is less an objective medical condition than something that exists in the eyes of a small band of influential psychiatrists. They legitimate their own interests by crafting an ideal script that seems beyond dispute, and they then labor to palm it off on practically everybody else. The proffering of an image of inherent order and incontrovertible normality, as wrongheaded as it is, is probably the major accomplishment of the *DSM*.

Mental disorders flesh out our understanding of the relativity of deviance. They show us that even private, personal experiences (such as "losing one's mind") are also public, social events. Power, conflict, and the social construction of reality all play a part in the manufacture of madness. An organized and influential group such as psychiatry has mustered its resources to craft understandings of sick and well that directly benefit it. Medicine generates meanings of the human condition that are given life through the words and deeds—the concrete actions and reactions—of people who use what they have learned to judge, evaluate, and classify certain acts and attributes as sick or abnormal.

Deviance is in the eyes of the beholders, it is true, but these beholders are not always lone moral crusaders. Each of us may be upset by what we believe are odd qualities of others—their appearances, eccentricities, body odors, and so on. However, more and more, what deviance is depends on the actions, interests, outlooks, and resources of representatives of formal organizations or professions (such as psychiatry) who can draw on a vast arsenal of meanings in their dealings with troublesome individuals. Institutions and formal organizations are important centers of social activity, and they provide rules and understandings about proper and improper ways of acting, thinking, feeling, and being.

References

American Psychiatric Association. (1987). *Diagnostic and statistical manual of mental disorders* (3rd ed., Rev.). Washington, DC: Author.

American Psychiatric Association. (1994). *Diagnostic and statistical manual of mental disorders* (4th ed.). Washington, DC: Author.

Armstrong, L. (1993). *And they call it help: The psychiatric policing of America's children*. Reading, MA: Addison-Wesley.

Banton, R., Clifford, P., Frosh, S., Lousada, J., & Rosenthall, J. (1985). *The politics of mental health*. New York: Macmillan.

Bartholomew, R. (1997). The medicalization of the exotic: *Latah* as a colonialism-bound "syndrome." *Deviant Behavior, 18,* 47-75.

Bayer, R. (1981). *Homosexuality and American psychiatry: The politics of diagnosis.* New York: Basic Books.

Breggin, P. (1991). *Toxic psychiatry: Why therapy, empathy, and love must replace the drugs, electroshock, and biochemical theories of the new psychiatry.* New York: St. Martin's.

Breggin, P., & Breggin, G. R. (1994). *The war against children.* New York: St. Martin's.

Caplan, P. (1995). *They say you're crazy: How the world's most powerful psychiatrists decide who's normal.* Reading, MA: Addison-Wesley.

Davies, V. (1947). *Miracle on 34th Street.* San Diego, CA: Harcourt Brace Jovanovich.

Dowbiggin, I. (1991). *Inheriting madness: Professionalization and psychiatric knowledge in nineteenth-century France.* Berkeley: University of California Press.

Fernando, S. (1991). *Mental health, race and culture.* New York: St. Martin's.

Figert, A. (1995). The three faces of PMS: The professional, gendered, and scientific structuring of a scientific disorder. *Social Problems, 42,* 56-73.

Francoeur, R. (1996). Introduction. In R. Francoeur (Ed.), *Taking sides: Clashing views on controversial issues in human sexuality* (5th ed., p. 44). Guilford, CT: Dushkin.

Frank, R. T. (1931). The hormonal causes of premenstrual tension. *Archives of Neurology and Psychiatry, 26,* 1053-1057.

Gallagher, B. J., III. (1995). *The sociology of mental illness* (3rd ed.). Englewood Cliffs, NJ: Prentice Hall.

Heinrichs, R. W. (1993). Schizophrenia and the brain: Conditions for a neuropsychology of madness. *American Psychologist, 48,* 221-233.

Herman, E. (1995). *The romance of American psychology: Political culture in the age of experts.* Berkeley: University of California Press.

Kirk, S., & Kutchins, H. (1992). *The selling of DSM: The rhetoric of science in psychiatry.* New York: Aldine de Gruyter.

Kirk, S., & Kutchins, H. (1994, June 20). Is bad writing a mental disorder? *New York Times,* p. A17.

Kovel, J. (1988). A critique of DSM-III. In S. Spitzer & A. Scull (Eds.), *Research in law, deviance and social control* (Vol. 9, pp. 127-146). Greenwich, CT: JAI.

Laws, S. (1983). The sexual politics of pre-menstrual tension. *Women's Studies International Forum, 6,* 19-31.

Martin, E. (1992). *The woman in the body: A cultural analysis of reproduction.* Boston: Beacon.

Mead, G. H. (1934). *Mind, self, and society: From the standpoint of a social behaviorist.* Chicago: University of Chicago Press.

Neill, J. (1995). "More than medical significance": LSD and American psychiatry 1953-1966. In J. Inciardi & K. McElrath (Eds.), *The American drug scene: An anthology* (pp. 214-220). Los Angeles, CA: Roxbury.

Rittenhouse, C. A. (1991). The emergence of premenstrual syndrome as a social problem. *Social Problems, 38,* 412-425.

Strauss, J. (1991). The meaning of schizophrenia: Compared to what? In W. Flack, Jr., D. Miller, & M. Wiener (Eds.), *What is schizophrenia?* (pp. 81-90). New York: Springer-Verlag.

Sumner, C. (1994). *The sociology of deviance: An obituary.* New York: Continuum.

Szasz, T. (1960). The myth of mental illness. *American Psychologist, 15,* 113-118.

Szasz, T. (1973). *The second sin.* Garden City, NY: Anchor.

Szasz, T. (1974). *The myth of mental illness* (Rev. ed.). New York: Harper & Row.

Szasz, T. (1985, September 28). Psychiatry: Rhetoric and reality. *Lancet, 2,* 711-712.

Szasz, T. (1997). Idleness and lawlessness in the therapeutic state. In L. Salinger (Ed.), *Deviant behavior 97/98* (2nd ed., pp. 214-219). Guilford, CT: Dushkin/Brown.

Tavris, C. (1996). The myth of PMS. In R. T. Francoeur (Ed.), *Taking sides: Clashing views on controversial issues in human sexuality* (5th ed., pp. 50-53). Guilford, CT: Dushkin/Brown & Benchmark.

Trent, J., Jr. (1994). *Inventing the feeble mind: A history of mental retardation in the United States.* Berkeley: University of California Press.

Turner, R., & Edgley, C. (1996). From witchcraft to drugcraft: Biochemistry as mythology. In H. Pontell (Ed.), *Social deviance: Readings in theory and research* (2nd ed., pp. 432-441). Upper Saddle River, NJ: Prentice Hall.

Umberson, D., Chen, M., House, J., Hopkins, K., & Slaten, E. (1996). The effect of social relationships on psychological well-being: Are men and women really so different? *American Sociological Review,* 837-857.

Wiener, M. (1991). Schizophrenia: A defective, deficient, disrupted, disorganized construct. In W. Flack, Jr., D. Miller, & M. Wiener (Eds.), *What is schizophrenia?* (pp. 199-222). New York: Springer-Verlag.

Afterword

Great diversity exists in human actions and attributes, and an even greater diversity exists in evaluations and judgments of them. Change is the only real constant, and social life is filled with conflicts, ambiguities, and absurdities. Whatever exists could be otherwise, and no matter how something is evaluated, it can always be viewed differently by some other group at another place and/or time. Practically anything that can be done with the human body has been done with the human body as a source of pleasure to somebody, somewhere, sometime; and practically any claim that can be made about what people have done with their bodies has been made. Humans have a remarkable capacity to make practically anything proper or improper, as the case may be.

Older conceptions of deviance and the deviant viewed them as having intrinsic qualities that objectively defined them. Deviance was usually viewed as something inherently bad (i.e., sick or harmful), and deviance was viewed either as what defective or sick people did or as what dysfunctional/disorganized societies produced. Although this approach still has its supporters (as we have seen), in sociology, attention has shifted to the role played by the "other" in the creation of deviance. Others can do things (or construct situations) that make it more likely that some people will become more engulfed in their deviant activities and more imprisoned by their stigmatizing attributes. Another thing that the "other" can do is to confer deviancy on some action or attribute. Once the ineradicable social nature of deviance was recognized, its relative and situational nature could never again be ignored. Explanations of deviance in terms of intrinsic and pathological factors became far less credible and convincing.

Social conflicts and different constructions of reality are at the very heart of the relativity of deviance. These social conflicts and different worldviews—we

189

saw them quite clearly in our examination of sexual diversity—become the raw stuff from which deviance emerges. It is even possible to be labeled deviant for what you are, which may cause stigma and isolation from others. A change in social relationships almost always produces a change in the forms that deviance takes. Even acts of "hard" deviance such as murder, rape, and suicide are not viewed the same way at all times and in all places. Just as no one trait characterizes all people who injure others, not all injury is defined as violence, and not all violence is defined as deviance. People who successfully offer motives and accounts that other people will accept may find that their actions and attributes are assessed in more positive ways. A corollary of this is that if one does *not* offer the proper motives and accounts, one's actions and attributes may be devalued more readily by others. Destructive acts may become so institutionalized that they actually become normative. Places may be found, for example, where raping women is expected, encouraged, or even trendy. Subcultures have their own momentum, and individuals may find that their use of interpersonal violence to get what they want is supported by others.

One of the most influential variables in the creation of social differentiation and social deviance is social power. Groups do what they do partly because of the amount of power that they have. One group may find itself doing things that its members would never do (or never be able to do) if they possessed a different amount of power. Beyond that, some groups will use their power to criminalize or to "deviantize" the activities of other groups. Moral entrepreneurs may decide that certain things ought not to happen and then set about to make things right. When moral crusades are successful and enduring, they create a whole new category of deviance and deviants. However, moral entrepreneurs rarely act as autonomous individuals. Usually, they are representatives of organizations, professions, agencies, or institutions (such as law, medicine, or religion) that benefit from the creation of deviance and the control or cure of deviants. Our exploration of the crafting of psychiatric nosology showed us that some groups will institutionalize and formalize understandings about proper/improper or healthy/sick that help them accomplish what they believe needs to be accomplished.

The development of a relativist perspective is an important achievement. It allows us to study how meanings, labels, and sanctions are constructed and applied to people in actual situations. Nothing is inherently deviant, and meanings are always problematic. The study of deviance drives home the point that the world contains a large number of social arrangements and an even larger number of individuals with unique characteristics and that some of these arrangements and individuals are defined as outside the pale by powerful groups in a socially constructed world.

Index

About the Author

John Curra is Full Professor of Sociology at Eastern Kentucky University, where he has taught since 1975. He received his master's degree from San Diego State University and his doctorate from Purdue University. He has taught courses in introductory sociology, social deviance, criminology, sociological analysis, social problems, social psychology, and juvenile delinquency. In 1981, he received the prestigious Excellence in Teaching Award from the College of Social and Behavioral Sciences. He is the author of *Understanding Social Deviance: From the Near Side to the Outer Limits* and an accompanying Instructor's Manual.